Handbook of
ELECTRONIC
CIRCUIT DESIGNS

Handbook of ELECTRONIC CIRCUIT DESIGNS

JOHN D. LENK
Consulting Technical Writer

Prentice-Hall, Inc., Englewood Cliffs, New Jersey

Library of Congress Cataloging in Publication Data

LENK, JOHN D (date)
   Handbook of electronic circuit designs.

   Includes index.
   1. Transistor circuits. 2. Electronic circuit
design. I. Title. II. Title: Electronic circuit
designs.
TK7871.9.L3639   621.3815'3'0422   75-8635
ISBN 0-13-377309-4

© 1976 by
PRENTICE-HALL, INC.
Englewood Cliffs, New Jersey

All rights reserved. No part of this book may be
reproduced in any form or by any means without
permission in writing from the publisher.

Printed in the United States of America

10 9 8 7 6 5 4 3 2 1

PRENTICE-HALL INTERNATIONAL, INC., *London*
PRENTICE-HALL OF AUSTRALIA, PTY. LTD., *Sydney*
PRENTICE-HALL OF CANADA, LTD., *Toronto*
PRENTICE-HALL OF INDIA PRIVATE LIMITED, *New Delhi*
PRENTICE-HALL OF JAPAN, INC., *Tokyo*
PRENTICE-HALL OF SOUTHEAST ASIA (PTE.) LTD., *Singapore*

To Irene, Mr. Lamb, the seagulls, and the sandpipers

CONTENTS

PREFACE   xi

1   FILTER AND ATTENUATOR CIRCUIT DESIGN   1

1-1. *RC* filters 2
1-2. *LC* filters 5
1-3. Active filters 17
1-4. Attenuators 20

2   RF CIRCUIT DESIGN CONSIDERATIONS   26

2-1. Resonant circuits for RF amplifiers 26
2-2. Basic RF amplifier design approaches 30
2-3. Types of RF amplifier tuning networks 48
2-4. Design of RF networks using
     voltage variable capacitors..................................... 54

3   WAVE-GENERATING CIRCUIT DESIGN   60

3-1. Sawtooth oscillators 60
3-2. Multivibrators 66
3-3. Schmitt triggers 72

4 PHOTOTRANSISTOR CIRCUITS 81

4-1. Phototransistor theory 81
4-2. Phototransistor static characteristics 82
4-3. Radiation and illumination sources 85
4-4. Low-frequency and steady-state design approaches for phototransistors 97
4-5. High-frequency design approaches for phototransistors 99

5 AF AMPLIFIER DESIGN EXAMPLES 107

5-1. The effect of amplifier components on frequency 107
5-2. Amplifier coupling circuit design 112
5-3. Amplifier design classifications 116
5-4. UJT regenerative amplifier 121
5-5. Basic two-junction transistor amplifier stage 123
5-6. Basic FET amplifier stage 127
5-7. Multistage transistor amplifiers 131
5-8. Direct-coupled transistor amplifiers 134
5-9. Multistage two-junction transistor amplifier with transformer coupling 146
5-10. Designing low-power audio amplifiers using plastic transistors 156

6 RF AMPLIFIER DESIGN EXAMPLES 170

6-1. RF voltage-amplifier design requirements 170
6-2. RF mixers and converters 174
6-3. AVC-AGC circuits for amplifiers 183
6-4. RF power amplifier and multiplier design requirements 184
6-5. RF amplifier design with datasheet graphs 194
6-6. RF voltage amplifier design 200

7 TRANSISTOR SWITCHES 207

7-1. Basic chopper circuits.. 207
7-2. Two-junction transistor choppers 209
7-3. FETs as choppers and switches 215
7-4. Transistor inverters and converters 235

8 OSCILLATOR CIRCUIT DESIGN   265

8-1. LC and crystal-controlled oscillators 265
8-2. RC oscillators 283
8-3. Blocking oscillators 291
8-4. Basic unijunction relaxation oscillator......................... 297

INDEX   303

PREFACE

The author's approach to electronic circuit design, which first appeared in his best seller *Handbook of Simplified Solid-State Circuit Design* (Prentice-Hall, Inc., Englewood Cliffs, N.J., 1971), is to start with approximations or guidelines for the selection of circuit component values on a trial basis, assuming a specified design goal and a given set of conditions. This handbook follows the same design approach and concentrates on simple, practical approaches to electronic circuit design, not on circuit analysis. Theory is included only where required for practical design.

With any electronic circuit it is possible to apply certain guidelines for the selection of component values. These guidelines can then be stated in basic equations requiring only simple arithmetic for their solution. The component values will depend upon transistor characteristics, available power source, desired performance (voltage amplification, stability, etc.), and external circuit conditions (input/output impedance match, input signal amplitude, etc.).

Transistor characteristics are to be found in the manufacturer's datasheets and other literature. The electronic circuit characteristics can be determined based on reasonable expectations of the transistor characteristics. Often, the final circuit is a result of many tradeoffs between desired performance and available characteristics. This handbook discusses the problem of tradeoffs from a simplified, practical standpoint.

It is assumed that the reader is already familiar with transistor basics (including two-junction transistors, FETs, and UJTs) at a level found in the author's *Handbook for Transistors* (Prentice-Hall, Inc., Englewood Cliffs, N.J., 1976). It is especially important that the reader be able to interpret transistor datasheets. However, *no direct reference to any of the author's previous books is required to understand and use this book.*

A wide range of electronic circuit designs is covered. A detailed listing can be found in the table of contents. In brief, this book covers the design of filters, attenuators, amplifiers (AF and RF), wave generators, phototransistor circuits, switching, and oscillator circuits.

Since this handbook does not require advanced math or theoretical study, it is ideal for the experimenter. On the other hand, this handbook is suited to schools where the basic teaching approach is circuit analysis, and a great desire exists for practical design.

The author has received much help from various individuals and organizations in writing this handbook. He wishes to give special thanks to the following: the Semiconductor Products Department of General Electric, the Semiconductor Products Division of Motorola Inc., the Components Group of Texas Instruments, and the Solid State Division of Radio Corporation of America. The author also wishes to thank Mr. Joseph A. Labok of Los Angeles Valley College.

JOHN D. LENK

1 FILTER AND ATTENUATOR CIRCUIT DESIGN

The primary purpose of a filter is to discriminate against the passage of certain groups of frequencies while simultaneously passing other groups or portions of the frequency spectrum. Although filter circuits range from the very simple to the very complex, there are only two basic types of filters, *active* and *passive*. The active filters described here are essentially a solid-state frequency-selective amplifier stage. The passive filters are either *RC* (resistor-capacitor), used primarily for low- or audio-frequency applications, or *LC* (coil-capacitor), for use at higher frequencies. Filters can also be classified by the frequencies that they are designed to pass or reject. The four most common filter classifications are as follows:

1. A *low-pass filter* is one that passes all frequencies below a selected value and attenuates higher frequencies. The low-pass filter is also known as a *high-cut filter*.
2. A *high-pass filter* is one that passes all frequencies above a selected value and attenuates lower frequencies. The high-pass filter is also known as a *low-cut filter*.
3. A *band-elimination filter*, also known as a band-stop, band-rejection, or band-suppression filter, is one that suppresses a selected band of frequencies while passing all lower and higher frequencies.
4. A *band-pass filter* is one that passes a selected band of frequencies while rejecting all lower and higher frequencies.

Like filters, *attenuators* function to diminish signal-strength level but not on a frequency-selective basis. Attenuators, as described here, are resistive networks designed to diminish signal strength and to match dissimilar impedances. The terms *attenuator* and *pad* are synonymous. However, a pad may or may not provide some attenuation, whereas an attenuator always provides some attenuation.

The attenuators and pads discussed here use fixed resistors and may be of two types: the *asymmetrical*, in which the input and output impedances are not the same, and the *symmetrical*, which have identical input and output impedances. In general, asymmetrical pads are used for impedance matching, but may also provide some attenuation. The symmetrical pads are used primarily as attenuators between two devices with equal impedances.

1-1. *RC* FILTERS

The use of resistance-capacitance (*RC*) filters is the simplest way to filter audio-frequency signals. Low frequencies are removed by a series capacitor and shunt resistor (Fig. 1-1), and high frequencies are filtered

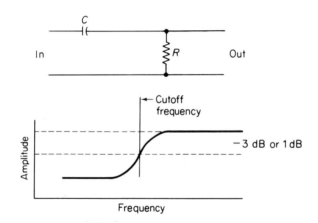

$$3 \text{ dB cutoff frequency} \approx \frac{1}{6.28RC}$$

$$3 \text{ dB } R \text{ (in ohms)} \approx \frac{1}{6.28FC}$$

$$3 \text{ dB } C \text{ (in farads)} \approx \frac{1}{6.28FR}$$

$$1 \text{ dB cutoff frequency} \approx \frac{1}{3.2RC}$$

$$1 \text{ dB } R \text{ (in ohms)} \approx \frac{1}{3.2FC}$$

$$1 \text{ dB } C \text{ (in farads)} \approx \frac{1}{3.2FR}$$

Figure 1-1 Basic high-pass (low-cut) *RC* filter

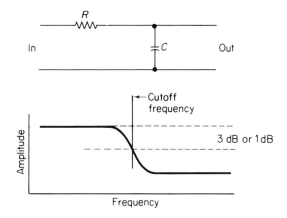

Figure 1-2 Basic low-pass (high-cut) *RC* filter

$$3 \text{ dB cutoff frequency} \approx \frac{1}{6.28RC}$$

$$3 \text{ dB } R \text{ (in ohms)} \approx \frac{1}{6.28FC}$$

$$3 \text{ dB } C \text{ (in farads)} \approx \frac{1}{6.28FR}$$

$$1 \text{ dB cutoff frequency} \approx \frac{1}{3.2RC}$$

$$1 \text{ dB } R \text{ (in ohms)} \approx \frac{1}{3.2FC}$$

$$1 \text{ dB } C \text{ (in farads)} \approx \frac{1}{3.2FR}$$

by interchanging the resistor and capacitor (Fig. 1-2). The filters in Figs. 1-1 and 1-2 are essentially *RC* circuits with the following three factors: resistance, reactance, and impedance. The amount of attenuation produced by *RC* filters is dependent on the ratio of resistance or of reactance to impedance.

In the high-pass (low-frequency attenuation) circuit in Fig. 1-1, input voltage is applied across both the resistor and the capacitor. Output voltage appears across the resistance. As frequency decreases, reactance increases, as does the impedance. Since the input voltage remains constant, the current through the circuit decreases ($I = E/Z$) with an increase in impedance. Therefore, the current through the resistance decreases and the output voltage drops. In a high-pass *RC* circuit, the attenuation is calculated by

$$\text{Output voltage} = \text{input voltage} \times \frac{R}{Z}$$

Since the value of R is fixed—for practical purposes—and the value of Z varies indirectly with frequency, the voltage output varies directly with frequency (output voltage drops with decreases in frequency).

In the low-pass (high-frequency attenuation) circuit in Fig. 1-2, the input voltage is applied across both the resistor and the capacitor. Output voltage is taken across the capacitor. As frequency increases, the reactance decreases, as does the impedance. However, since the impedance is composed of both reactance and fixed resistance, the impedance does not decrease as much as the reactance. Therefore, the decrease in reactance (in relation to impedance) causes the voltage output to decrease. In a low-pass RC filter circuit, the attenuation is calculated by:

$$\text{Output voltage} = \text{input voltage} \times \frac{X_C}{Z}$$

Since the reactance varies at a greater rate than the impedance, the voltage output varies directly with frequency (voltage drops with increases in frequency).

1-1.1. Design Considerations for RC Filters

Filter attenuation is rated in terms of decibel drop at a given frequency. Generally, RC filters are designed to produce an approximate 3-dB drop (to 0.707 of input) at a selected cutoff frequency. However, some RC filters are designed to produce a 1-dB drop at the cutoff frequency. The equations for both the 1-dB and 3-dB drops are given in Figs. 1-1 and 1-2.

In any RC filter, either the capacitor or the resistor value could be assumed to find the other value for a given cutoff frequency. In practical design the resistance value is usually assumed, because the resistor is chosen to meet the other circuit requirements. For example, the resistance in a high-pass filter may also form the circuit's input or output impedance.

1-1.2. Band-Pass RC Filters

Both the high-pass (low-cut) and low-pass (high-cut) filter circuits can be combined to provide a band-pass RC filter, as shown in Fig. 1-3. As a guideline, the high-frequency limit (F_H) must be at least 10 times the low-frequency limit (F_L) for such a band-pass circuit to be effective. If the high- and low-frequency limits are not greater than 10 to 1, there will be considerable interaction between the combined circuits. Even with a 10-to-1 ratio, some interaction may occur. Thus, the equations in Fig. 1-3 are approximate.

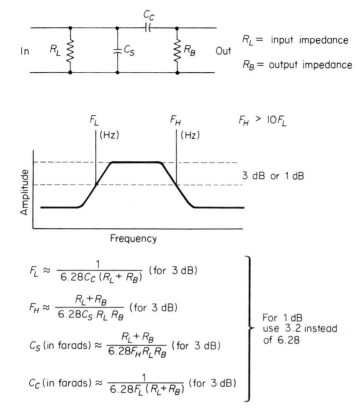

Figure 1-3 Basic band-pass RC filter

$$F_L \approx \frac{1}{6.28 C_C (R_L + R_B)} \text{ (for 3 dB)}$$

$$F_H \approx \frac{R_L + R_B}{6.28 C_S R_L R_B} \text{ (for 3 dB)}$$

$$C_S \text{ (in farads)} \approx \frac{R_L + R_B}{6.28 F_H R_L R_B} \text{ (for 3 dB)}$$

$$C_C \text{ (in farads)} \approx \frac{1}{6.28 F_L (R_L + R_B)} \text{ (for 3 dB)}$$

For 1 dB use 3.2 instead of 6.28

1-1.3. Multistage RC Filters

A single *RC* filter will provide a gradual transition from the passband to the cutoff region. If a rapid transition is necessary for design, two or more *RC* filter stages can be combined, as shown in Fig. 1-4. The increase in attenuation at the cutoff frequency, and at frequencies above and below the cutoff point, for a two-stage high-pass filter are shown in Fig. 1-5. A similar curve for a two-stage low-pass filter is shown in Fig. 1-6. Any number of stages may be added to an *RC* filter. (As a guideline, each stage will increase the attenuation by 6 dB at the cutoff frequency.)

1-2. LC FILTERS

The most precise filters are composed of inductors (coils) and capacitors. An *LC* filter can also include resistors. The input and output

6 Filter and Attenuator Circuit Design   Chap. 1

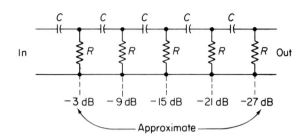

$$\text{Cutoff frequency} \approx \frac{1}{6.28RC}$$

$$R \text{ (in ohms)} \approx \frac{1}{6.28FC}$$

$$C \text{ (in farads)} \approx \frac{1}{6.28FR}$$

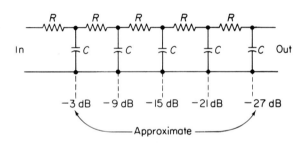

Figure 1-4 Basic multistage RC filters

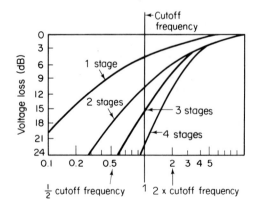

Figure 1-5 Voltage loss curve for high-pass RC filter stages

Sec. 1-2　　　　　　　　　　　　　　　　　　　LC Filters　　7

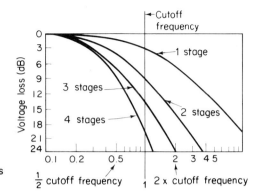

Figure 1-6　Voltage loss curves for low-pass RC filter stages

connections of most LC filters are terminated in source and load resistances, or by impedances equal in value to the impedance of the filter. Except for special applications, it is not practical to use LC filters for audio frequencies. The inductances required at low frequencies are quite large, and therefore heavy and bulky.

1-2.1. Low-Pass LC Filter

Figure 1-7 shows a typical low-pass LC filter in its simplest form, the *basic L configuration*. All LC filters take advantage of the fact that capacitors and coils operate inversely in the presence of alternating current. That is, inductive reactance increases with frequency, and capacitive reactance

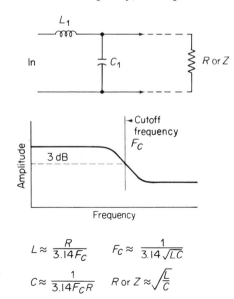

Figure 1-7　Basic L low-pass constant-K filter

$$L \approx \frac{R}{3.14 F_C} \qquad F_C \approx \frac{1}{3.14 \sqrt{LC}}$$

$$C \approx \frac{1}{3.14 F_C R} \qquad R \text{ or } Z \approx \sqrt{\frac{L}{C}}$$

decreases with frequency. Thus in the low-pass LC filter, the parallel unit (a capacitor across the line) has a decreasing reactance as frequency is increased. This acts to bypass high frequencies, but has an increasing reactance to low frequencies. The series unit (a coil in the line) has an increasing reactance to high frequencies but passes the low frequencies.

In most LC filters, the product of the impedances presented by the capacitance and inductance remains constant with changes in frequency (because of the inverse change in reactance). For example, if capacitive reactance goes down for a given increase in frequency, the inductive reactance goes up by a corresponding amount. The characteristic impedance of the filter remains constant—hence the name *constant-K filter*.

One of the problems with the basic L-type filter is that it does not provide sharp cutoff frequency (f_c). To increase cutoff frequency sharpness, another coil can be added to the basic L configuration, as shown in Fig. 1-8. Such a filter is known as the *T-type* because of its appearance. In a T filter, the value of capacitor C does not change from that of a basic L configuration, and the basic equations are the same. The total inductance of L_1 and L_2 must be equivalent to that of the single coil in a basic L configuration. Usually the desired total inductance is divided evenly between the two coils, so each coil in a T-type low-pass filter has *one-half the total desired inductance*.

Frequency sharpness can also be increased by adding another capacitor,

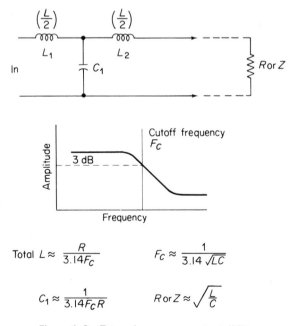

Figure 1-8 *T*-type low-pass constant-K filter

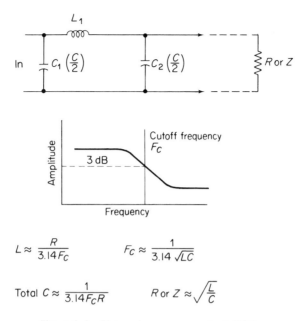

Figure 1-9 Pi-type low-pass constant-K filter

as shown in Fig. 1-9. Such a filter is known as the *Pi-type*. In a Pi-type LC filter, the value of L does not change, but the total capacitance of C_1 and C_2 must be equivalent to that of the single capacitor in a basic L configuration. Usually the desired total capacitance is divided evenly between the two capacitors, so each capacitor in a Pi-type low-pass filter has *one-half the total desired capacitance.*

1-2.2. High-Pass LC Filter

Figure 1-10 shows a typical high-pass filter in its simplest form, the basic L configuration. High-pass filters also take advantage of the fact that the capacitors and coils operate inversely in the presence of alternating current. Thus, in a high-pass LC filter, the series unit (a capacitor in the line) has a decreasing reactance as the frequency is increased. The series unit passes high frequencies but offers an increasing reactance to low frequencies. The parallel unit (a coil across the line) bypasses low frequencies but has an increasing reactance to high frequencies. Most high-pass LC filters are of the constant-K type described in Sec. 1-2.1.

To increase cutoff frequency (f_c) sharpness, another capacitor can be added to the basic L configuration, as shown in Fig. 1-11. Such a filter is known as the *T*-type. In the T filter, the value of coil L does not change from that of the basic L configuration, and the basic equations are the same. The

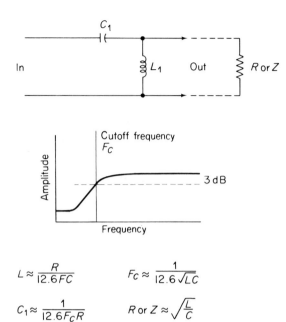

Figure 1-10 Basic L high-pass constant-K filter

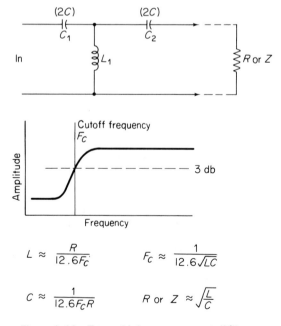

Figure 1-11 T-type high-pass constant-K filter

total capacitance of C_1 and C_2 must be equivalent to that of the single capacitor in the basic L configuration. Usually the desired total capacitance is divided evenly between the two capacitors, so each capacitor in the T-type high-pass LC filter has *twice the total desired capacitance*.

Frequency sharpness can also be increased by adding another coil, as shown in Fig. 1-12. Such a filter is known as the Pi-type. In a Pi-type LC filter, the value of C does not change, but the total inductance of L_1 and L_2 must be equivalent to that of the single coil in a basic L configuration. Usually the desired total inductance is divided evenly between the two coils, so each coil in the Pi-type high-pass LC filter has *twice the total desired inductance*.

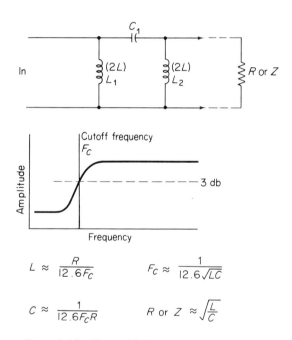

Figure 1-12 Pi-type high-pass constant-K filter

1-2.3. Band-Elimination LC Filter

The band-elimination filter takes advantage of the different impedance characteristics of series- and parallel-resonant LC circuits. A parallel LC circuit has maximum impedance at the resonant frequency. A series LC circuit has minimum impedance at the resonant frequency.

These two LC circuits are combined in the band-elimination filter circuit in Fig. 1-13. The series arm has minimum impedance at the center frequency

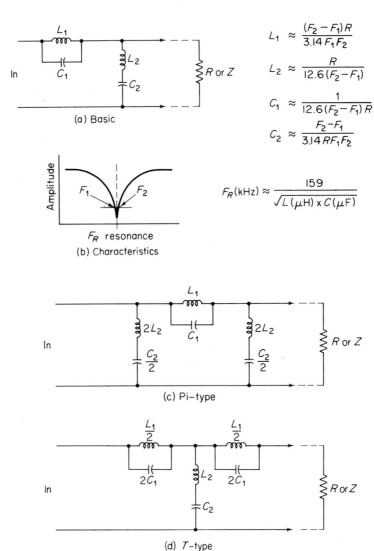

Figure 1-13 Band-elimination filter circuits

of the desired band. The impedance increases on either side of resonance. This acts to bypass the center frequency.

The parallel arm has maximum impedance at the center frequency, with the impedance decreasing on either side of resonance. This prevents passage of the center frequency as well as a band of frequencies on either side.

1-2.4. Band-Pass LC Filter

The band-pass filter takes advantage of the differing impedance characteristics of series- and parallel-resonant circuits. A parallel LC circuit has maximum impedance at the resonant frequency. A series LC circuit has minimum impedance at the resonant frequency.

These two LC circuits are combined in the band-pass filter in Fig. 1-14. The series arm has minimum impedance at the center frequency of the

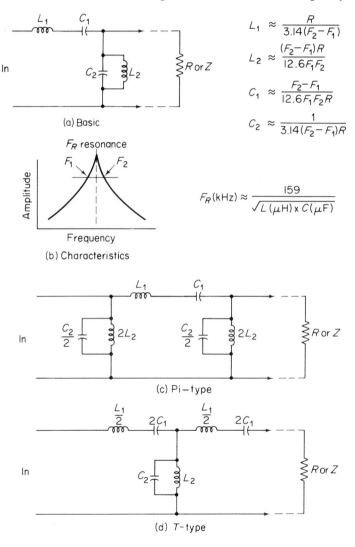

Figure 1-14 Band-pass filter circuits

desired band. The impedance increases on either side of resonance. This acts to pass the center frequency and to reject frequencies above and below the center frequency.

The parallel arm has maximum impedance at the center frequency, and the impedance decreases on either side of resonance. This acts to bypass frequencies above and below the center. Therefore, both the series and parallel arms provide passage of the center frequency as well as a band of frequencies on either side of center.

1-2.5. Low-Pass m-Derived LC Filter

When a sharper and more defined cutoff point is required than can be obtained from the constant-K filter, an m-derived filter is used. The m-derived filter is a basic constant-K type filter with the addition of another

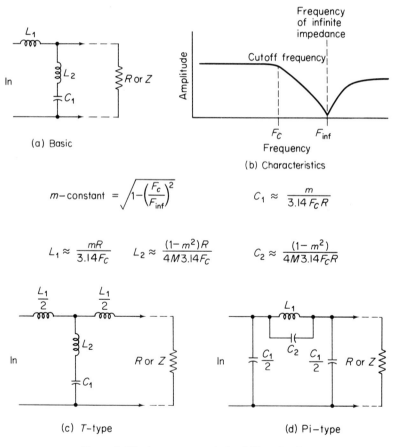

Figure 1-15 Low-pass m-derived filter circuits

component in series or shunt. This is shown in Fig. 1-15(a), where inductance L_2 is added to the original half-section low-pass filter as in Fig. 1-7. In the circuit of Fig. 1-15(a), C_1 and L_2 will be resonant for a particular frequency and will provide a low-impedance shunt. The m-derived filter is so designed that infinite attenuation is obtained at a specific frequency beyond cutoff frequency (f_c). The component impedances are interrelated by the m-constant which, in equation form, is related to the ratio of the cutoff frequency (f_c) and the frequency of infinite attenuation (f_{inf}). The m-constant is a fractional number between 0 and 1 and usually has an approximate value of 0.6. For a more pronounced cutoff, the m-constant is set nearer to 0.

1-2.6. High-Pass m-Derived LC Filter

As discussed for the low-pass filter, m-derived sections are also used in place of the high-pass constant-K types when a sharper and more defined cutoff point is required. The additional components of C_2 and L_2 are shown in Fig. 1-16.

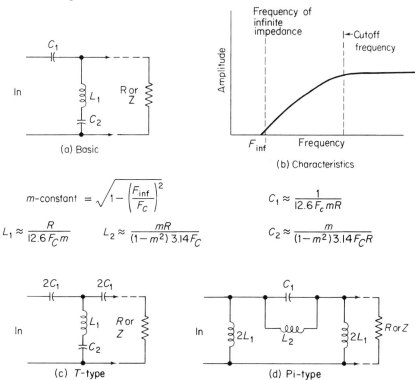

Figure 1-16 High-pass m-derived filter circuits

1-2.7. Frequency Calculations for m-Derived LC Filters

Unlike the equations used to calculate the frequency in the simple constant-K filters, the equations used to calculate the frequency in m-derived filters are different for shunt-derived and series-derived. When the reactance is added to the series arm, the filter is said to be *shunt-derived*. When the reactance is added to the shunt arm, the filter is called *series-derived*.

The equations for frequency calculation of various m-derived filter circuits are given in Figs. 1-17 through 1-20.

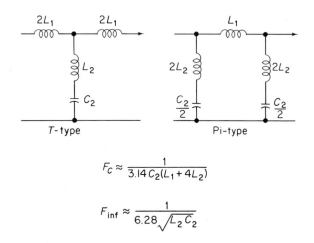

$$F_c \approx \frac{1}{3.14\, C_2(L_1 + 4L_2)}$$

$$F_{inf} \approx \frac{1}{6.28\sqrt{L_2 C_2}}$$

Figure 1-17 Low-pass series-derived filter circuits

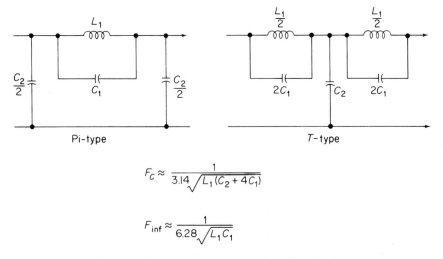

$$F_c \approx \frac{1}{3.14\sqrt{L_1(C_2 + 4C_1)}}$$

$$F_{inf} \approx \frac{1}{6.28\sqrt{L_1 C_1}}$$

Figure 1-18 Low-pass shunt-derived filter circuits

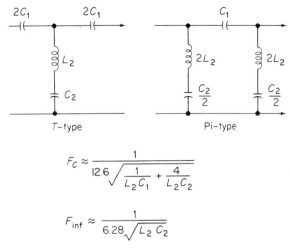

Figure 1-19 High-pass series-derived filter circuits

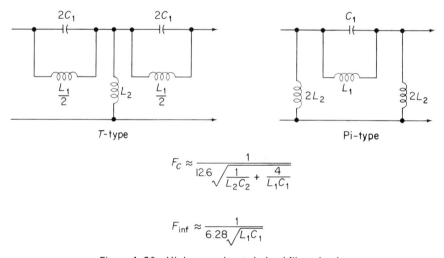

Figure 1-20 High-pass shunt-derived filter circuits

1-3. ACTIVE FILTERS

In addition to passive filters (either LC or RC), it is possible to use amplifiers to form active filters. There are two prime advantages in using these active filters. First, it is possible to obtain the equivalent of an inductive reactance without actually using a heavy and bulky inductance usually required for a typical LC audio filter. (LC filters are generally not practical in the audio-frequency range.) Second, the use of an active filter eliminates the signal loss normally associated with passive filters (either RC or LC).

1-3.1. Active Low-Pass (High-Cut) Filter

Figure 1-21 shows the basic circuit of an active low-pass filter together with the corresponding characteristic curves for several sets of component values. Note that these values are approximate and will usually require trimming to achieve an exact curve. The typical voltage gain is slightly less than 1 (unity) for transistors with a minimum β of 20.

The amount of gain and the shape of the curves are set by the amount of feedback in relation to signal (which, in turn, is set by component values). Note that the feedback is positive, and thus adds to the signal. However, the feedback amplitude (across the entire frequency range) is just below the point necessary for oscillation. The circuit in Fig. 1-21 is an emitter follower, which typically has no voltage gain.

The circuit in Fig. 1-21 requires a bias of approximately -10 V (one-half

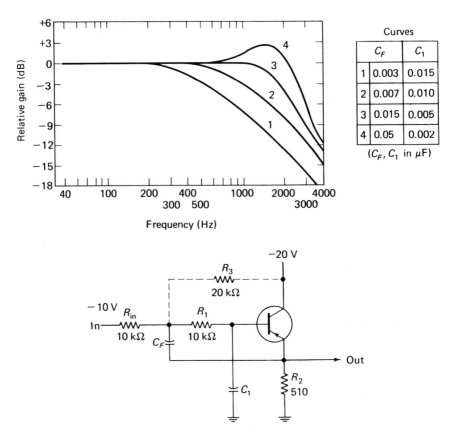

Figure 1-21 Active low-pass (high-cut) filter and corresponding response curves

Curves	C_F	C_1
1	0.003	0.015
2	0.007	0.010
3	0.015	0.005
4	0.05	0.002

(C_F, C_1 in μF)

the -20-V supply) at the input. This can be obtained from a previous stage. If no such stage exists, the bias can be obtained by the addition of the 20-kΩ resistor (shown in phantom as R_3) and by changing the value of R_{IN} to 20 kΩ. Such an arrangement will introduce a loss of about 6 dB. Thus, it is better to operate the circuit in Fig. 1-21 by direct coupling from the output of a previous stage.

1-3.2. Active High-Pass (Low-Cut) Filter

Figure 1-22 shows the basic circuit of an active high-pass filter, together with characteristic curves. The circuit of Fig. 1-22 is the inverse of

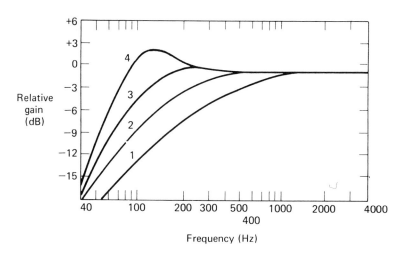

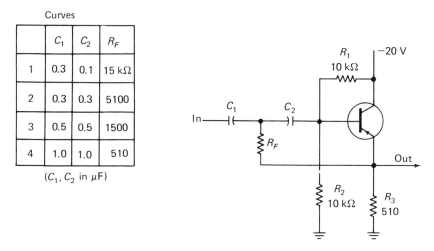

Curves

	C_1	C_2	R_F
1	0.3	0.1	15 kΩ
2	0.3	0.3	5100
3	0.5	0.5	1500
4	1.0	1.0	510

(C_1, C_2 in μF)

Figure 1-22 Active high-pass (low-cut) filter and corresponding curves

the Fig. 1-21 circuit. That is, the Fig. 1-22 circuit used capacitors in series with the base, with feedback obtained through R_F rather than C_F. The gain and shape of the curves are set by the amount of feedback (determined by circuit values).

1-3.3. Active Band-Pass Filter

The circuits in Figs. 1-21 and 1-22 can be cascaded to provide a band-pass filter. Any of the curves can be used. However, curve 3 is the most satisfactory because it has the sharpest break at cutoff. Curves 1 and 2 have considerable slope with no sharp break, and curve 4 produces some peaking at the breakpoints.

If the circuits are cascaded, the low-pass filter (Fig. 1-21) should follow the high-pass filter (Fig. 1-22). This provides the necessary bias at the input of the low-pass filter (-10 V from the emitter of the high-pass filter).

1-3.4. Active Peaking Filter

The circuits of Figs. 1-21 and 1-22 can be cascaded to provide a peaking filter with the proper selection of components. However, a single-stage tuned amplifier will produce the same results. Such a circuit is shown in Fig. 1-23.

As shown by the characteristic curve, the center, or peak frequency, is approximately 1 kHz. If desired, the center frequency can be changed by as much as 3 decades when *both capacitors are changed by a common factor*. However, in a practical circuit, the input resistance values will require some trimming.

1-4. ATTENUATORS

Although there are any number of attenuator, or pad, circuits, they are generally versions of a few basic types. The basic types include the L-, U-, T-, O-, H-, and Pi-types. In general, the L- and U-pads are assymetrical (for impedance matching), and the T-, O-, H-, and Pi-types are symmetrical (for attenuation).

1-4.1. Asymmetrical L- and O-Pads

The basic L-type pad shown in Fig. 1-24 is an asymmetrical pad used to match impedance of an input source to that of an output source. This provides a smooth transition between devices of different characteristic impedances, with each device encountering a total impedance equal to its own characteristic impedance. Although a certain amount of signal attenua-

Sec. 1-4 Attenuators

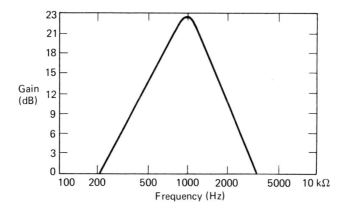

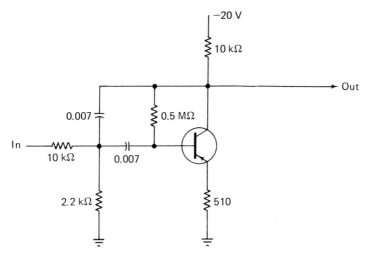

Figure 1-23 Active peaking filter and corresponding response curve

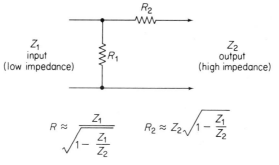

$$R \approx \frac{Z_1}{\sqrt{1 - \frac{Z_1}{Z_2}}} \qquad R_2 \approx Z_2\sqrt{1 - \frac{Z_1}{Z_2}}$$

Figure 1-24 Basic L-type pad and calculations

tion is usually required to make this transition, the pad in Fig. 1-24 will introduce a minimum of loss while performing its impedance-matching function.

If Z_1 (input) has a lower impedance with respect to Z_2 (output), the relationships and equations are as shown in Fig. 1-24. In practice, the standard resistor values that are closest to the calculated values are used for R_1 and R_2.

A balanced arrangement of the L-type pad is shown in Fig. 1-25. Here, each series resistor has one-half the value of R_2, shown originally in Fig. 1-24. When the pads are placed in cascade, as in Fig. 1-25(b), the network is sometimes referred to as a *ladder pad*. The basic pad of Fig. 1-25 is often known as a U-pad (because of its shape).

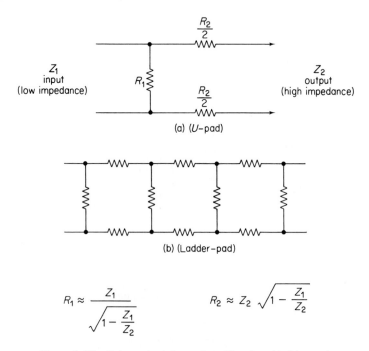

Figure 1-25 Balanced minimum-loss U-pad and ladder-pad

1-4.2. Symmetrical T- and H-Pads (Attenuators)

The T-pad, shown in Fig. 1-26, is a symmetrical pad where the impedance of the input device matches that of the output device. The only function of the T-pad is *attenuation of the signal*. Since impedance matching is not involved, the values of all R_1 resistors are identical, and the values of

R_1 and R_2 are chosen to provide the degree of attenuation desired. The T-pad in Fig. 1-26(a) is an unbalanced type. The balanced version (sometimes known as an H-pad) is shown in Fig. 1-26(b). In the balanced version, the R_1 values are cut in half.

The values of R_1 and R_2 are related to the ratio of signal-voltage or signal-current attenuation that is desired. It is usually simpler to use a voltage ratio. For example, if a signal voltage having an amplitude of 100 V is to be attenuated to produce a 10-V output signal, the voltage ratio is 10. If the desired attenuation is expressed in decibels, the value can be converted to a voltage ratio.

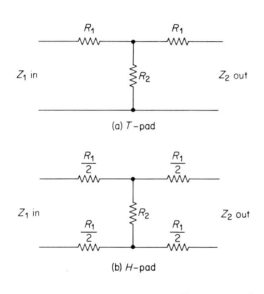

$$R_1 \approx Z \frac{(V-1)}{(V+1)} \qquad R_2 \approx Z \left[\frac{2V}{(V+1)(V-1)} \right]$$

V = input voltage divided by output voltage

Figure 1-26 T-pad and H-pad circuits for attenuation

1-4.3. Symmetrical Pi- and O-Pads (Attenuators)

The Pi-type pad shown in Fig. 1-27(a) is a symmetrical unbalanced type. The balanced version (sometimes known as an O-pad) is shown in Fig. 1-27(b). Since the input impedance is the same as the output impedance, no impedance matching is involved. The resistor values are chosen to provide the degree of attenuation desired. As with the T-pad and H-pad, the equations for solving the resistor values must consider the voltage attenuation ratio.

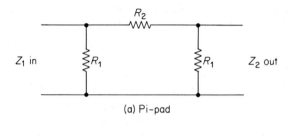

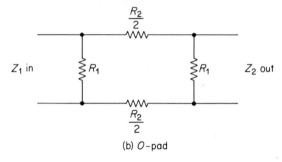

$$R_1 \approx Z \frac{(V+1)}{(V-1)} \qquad R_2 \approx Z\left(\frac{V^2-1}{2V}\right)$$

V = input voltage divided by output voltage

Figure 1-27 Pi-pad and O-pad circuits for attenuation

1-4.4. Symmetrical Bridged T- and H-Pads

An additional resistor can be shunted across the series resistors of the T-pad and H-pad to form what is known as a *bridged pad*. The bridged T- and H-pads are shown in Figs. 1-28(a) and 1-28(b), respectively. The ohmic values of R_1 and R_2 are chosen so each has a resistance equal to the ohmic value of the impedance; therefore only the resistance of R_3 and R_4 need be calculated.

Sec. 1-4   Attenuators

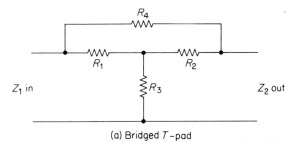

(a) Bridged T-pad

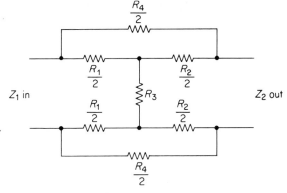

(b) Bridged H-pad

$$R_1 = R_2 = Z \qquad R_3 \approx \frac{Z}{V-1} \qquad R_4 \approx Z(V-1)$$

V = input voltage divided by output voltage

Figure 1-28   Bridged T-pad and bridged H-pad circuits

2 RF CIRCUIT DESIGN CONSIDERATIONS

Two-junction transistors and FETs are both used in RF circuit design. Generally, FETs are limited to oscillator and voltage-amplifier applications. Two-junction transistors can be used as oscillators and in either voltage-amplifier or power-amplifier circuits. Practically any of the traditional vacuum tube RF circuits can be duplicated with transistors.

The most important information regarding RF circuit considerations is the design of the tuning networks. These networks not only provide tuning to the desired operating frequency, but also match transistor characteristics to input and output circuit impedances. For these reasons, we shall concentrate on problems associated with these networks in this chapter.

The specific design problems for RF amplifier circuits are discussed in Chapter 6. The design considerations for oscillator circuits, including RF oscillators, are covered in Chapter 8.

2-1. RESONANT CIRCUITS FOR RF AMPLIFIERS

RF amplifier design is based on the use of resonant circuits (tank circuits) consisting of a capacitor and a coil (inductance) connected in series or parallel, as shown in Fig. 2-1. At the resonant frequency, the inductive and capacitive reactances are equal, and the circuit acts as a high impedance (if it is a parallel circuit) or a low impedance (if it is a series circuit). In either case, any combination of capacitance and inductance has some resonant frequency.

Either (or both) the capacitance or inductance can be variable to permit

Capacitive Reactance

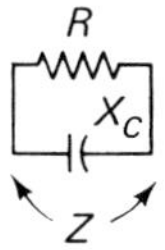

Parallel:

$$Z = \frac{RX_C}{\sqrt{R^2 + X_C^2}} \qquad Q = \frac{R}{X_C}$$

Series:

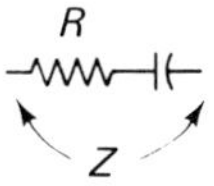

$$Z = \sqrt{R^2 + X_C^2} \qquad Q = \frac{X_C}{R}$$

$$X_C = \frac{159}{F\text{ (kHz)} \times C\text{ (}\mu\text{F)}}$$

$$F\text{ (kHz)} = \frac{159}{(X_C) \times C\text{ (}\mu\text{F)}}$$

$$C\text{ (}\mu\text{F)} = \frac{159}{F\text{ (kHz)} \times X_C}$$

Inductive Reactance

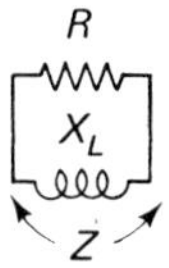

Parallel:

$$Z = \frac{RX_L}{\sqrt{R^2 + X_L^2}} \qquad Q = \frac{R}{X_L}$$

Series:

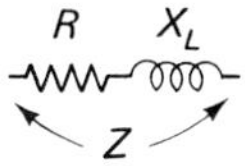

$$Z = \sqrt{R^2 + X_L^2} \qquad Q = \frac{X_L}{R}$$

$$X_L = 6.28 \times F\text{ (kHz)} \times L\text{ (mH)}$$

$$F = \frac{X_L}{6.28L}$$

$$L = \frac{X_L}{6.28F}$$

Impedance and Resonance

Series (zero impedance)

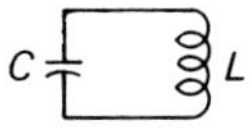

Parallel (infinite impedance*)

$$F\text{ (kHz)} = \frac{159}{\sqrt{L\text{ (}\mu\text{H)} \times C\text{ (}\mu\text{F)}}}$$

$$L\text{ (}\mu\text{H)} = \frac{2.54 \times 10^4}{F\text{ (kHz)}^2 \times C\text{ (}\mu\text{F)}}$$

$$C\text{ (}\mu\text{F)} = \frac{2.54 \times 10^4}{F\text{ (kHz)}^2 \times L\text{ (}\mu\text{H)}}$$

*When circuit Q is 10 or higher

Figure 2-1 Resonant circuit equations

tuning of the resonant circuit over a given frequency range. When the inductance is variable, the circuit is usually tuned by means of a metal slug (usually powdered iron) inside the coil. The metal slug is screwdriver adjusted to change the inductance (and thus the inductive reactance) as required.

Typical RF circuits used in receivers (AM, FM, communications, etc.) often include two resonant circuits in the form of a transformer (RF or IF transformer, etc.). Either the capacitance or inductance can be variable.

In the case of RF transmitter circuits, it is sometimes necessary to design the coil portion of the resonant circuit. This is because coils of a given inductance and physical size may not be available from commercial sources.

2-1.1. Basic Design Considerations for Resonant Circuits

The two most important considerations for RF resonant circuits are *resonant frequency* and the Q (or *quality* factor).

Resonant frequency. Figure 2-1 contains equations which show the relationship between capacitance, inductance, reactance, and frequency as they relate to resonant circuits. Note that there are three sets of equations. Two sets include reactance (inductive and capacitive), and the third set omits reactance. The reason for three sets of equations is that some design approaches require the reactance to be calculated for resonant networks. Solid-state RF transmitter circuits are a good example of this.

Quality factor and selectivity. A resonant circuit has a Q, or quality, factor which is directly related to the selectivity of the circuit, and is dependent upon the ratio of reactance to resistance. If a resonant circuit has pure reactance, the Q is high (theoretically infinite). However, this is not practical. For example, any coil and the leads of a capacitor will have some d-c resistance. Also, as frequency increases, the a-c resistance presented by the leads will increase due to skin effect. The sum total of these resistances is usually lumped together and considered as a resistor in series or parallel with the resonant circuit. The total resistance is usually termed the *effective* resistance, and is not to be confused with the reactance.

The resonant circuit Q is dependent upon the individual Q factors of both the inductance and capacitance used in the circuit. For example, if both the inductance and capacitance have a high Q, the circuit will have a high Q, provided that a minimum of resistance is produced when the inductance and capacitance are connected to form a resonant circuit.

Usually, resonant circuit Q is measured at points on either side of the resonant frequency where the signal amplitude is down 0.707 of the peak resonant value, as shown in Fig. 2-2. Notice that the resonant circuit with a high Q produces a sharp resonance curve (narrow bandwidth), whereas a low Q produces a broad resonance curve (wide bandwidth). For example, a high

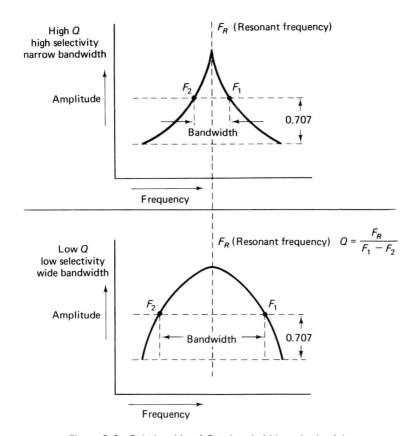

Figure 2-2 Relationship of Q to bandwidth and selectivity

Q resonant circuit will provide good harmonic rejection (tend to pass only the fundamental frequency) and efficiency in comparison with a low Q circuit, all other factors being equal. Thus, the *selectivity* of a resonant circuit is related directly to the Q.

A very high Q (or high selectivity) is not always desired. In some applications, it is necessary to add resistance to a resonant circuit to broaden the response (increase the bandwidth, decrease the selectivity). An example of this is the damping resistor used across peaking coils in a video amplifier.

If a given bandwidth must be maintained, but the resonant frequency is increased, the Q must also increase. For example, if the resonant frequency is 30 MHz with a bandwidth of 3 MHz, the required circuit Q is 10. If the resonant frequency is increased to 54 MHz with the same 3-MHz bandwidth, the required Q is 18. Also, Q must be decreased for increases in bandwidth, if the same resonant frequency is to be maintained.

2-1.2. Basic Design Considerations for RF Coils

The equations necessary to calculate the self-inductance of a single-layer air-core coil are given in Fig. 2-3. This type of coil is most efficient (maximum inductance for minimum physical size) when the ratio of coil radius to coil length is 1.25, that is, when the length is 0.8 of the radius.

The equations of Fig. 2-3 are approximations only, and do not take into account such factors as uneven sizes of turns, spacing between the turns, and the like. From a practical standpoint, use the equations to find the nearest number of turns (for a given inductance), and then spread or compress the turns as necessary to obtain a precise value of inductance (as measured on an inductance bridge).

2-2. BASIC RF AMPLIFIER DESIGN APPROACHES

Solid-state RF amplifiers can be designed using *two-port networks*. Basically, the method consists of characterizing the transistor as a linear active two-port network (LAN) with admittances (y-parameters), and using the parameters to solve design equations for stability, gain, and input-output admittances. The two-port design approach is best suited for *voltage amplifiers* using *small-signal* characteristics or parameters. The two-port

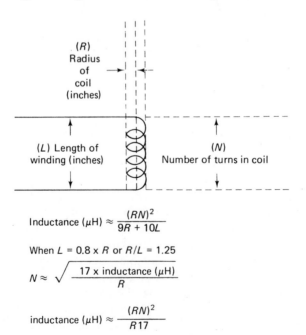

$$\text{Inductance } (\mu H) \approx \frac{(RN)^2}{9R + 10L}$$

When $L = 0.8 \times R$ or $R/L = 1.25$

$$N \approx \sqrt{\frac{17 \times \text{inductance } (\mu H)}{R}}$$

$$\text{inductance } (\mu H) \approx \frac{(RN)^2}{R17}$$

Figure 2-3 Calculations for self-inductance of single-layer air-core coil

system is recommended for all FET RF amplifiers, and for low-power two-junction RF amplifiers. Design based on *large-signal* characteristics (transistor input/output resistances and capacitance) is recommended for RF power amplifiers (using high-power two-junction transistors). Both design approaches are discussed in this chapter and in Chapter 6.

With either approach, it is difficult at best to provide simple, step-by-step procedures for designing RF amplifiers to meet all possible circuit conditions. In practice, there are several reasons why this procedure often results in considerable trial and error.

First, not all of the characteristics are always available in datasheet form. For example, input and output admittances may be given at some low frequency, but not at the desired operating frequency.

Often, manufacturers do not agree on terminology. A good example of this is in y-parameters, where one manufacturer uses letter subscripts (y_{fs}) and another uses number subscripts (y_{21}). Of course, this type of variation can be eliminated by conversions.

In some cases, manufacturers will give the required information on datasheets, but not in the required form. For example, some manufacturers may give the input capacitance in farads rather than listing the input admittance in mhos. The input admittance is found when the input capacitance is multiplied by $6.28F$ (where F is the frequency of interest). This is based on the assumption that the input admittance is primarily capacitive, and thus dependent on frequency. The assumption is not always true for the frequency of interest; therefore, it may be necessary to use complex admittance measuring equipment to make actual tests of the transistor.

The input and output tuning circuits of an RF amplifier must perform three functions. Obviously, the circuits (capacitors and coils) must tune the amplifier to the desired frequency. In addition, the circuits must match the input and output impedances of the transistor to the impedances of the source and load; otherwise, there will be considerable loss of signal. Finally, as in the case with any amplifier, there is some feedback between output and input. If the admittance factors are just right, the feedback will be of sufficient amplitude and of proper phase to cause oscillation of the amplifier. The amplifier is considered as *unstable* when this occurs.

Amplifier instability in any form is always undesirable, and can be corrected by feedback (called *neutralization*) or by changes in the input-output tuning networks. Although the neutralization and tuning circuits are relatively simple, the equations for determining stability (or instability) and impedance matching are long and complex. Generally, such equations are best solved by computer-aided design methods.

In an effort to cut through this maze of information and complex equations, we shall discuss all of the steps involved in RF amplifier design. Armed with this information, the reader should be able to interpret datasheets or test information, and use the information to design tuning networks that will provide stable RF amplification at the frequencies of interest. With each step we shall discuss the various alternative procedures and types of information available. Specific design examples of RF amplifier networks (given in Chapter 6) summarize the information contained in this chapter. On the assumption that not all readers are familiar with two-port networks, we shall start with a summary of the y-parameter system.

2-2.1. y-Parameters

Impedance (Z) is a combination of resistance (the real part) and reactance (the imaginary part). Admittance (y) is the reciprocal of impedance, and is composed of conductance (the real part) and susceptance (the imaginary part). A y-parameter is an expression for admittance in the form

$$y_{is} = g_{is} + jb_{is}$$

where g_{is} is the real (conductive) part of common-source input admittance, jb_{is} is the imaginary (susceptive) part of input admittance, and y_{is} is simply the input admittance.

The *term* $y_{is} = g_{is} + jb_{is}$ expresses the y-parameter in *rectangular form*. Some manufacturers describe the y-parameter in *polar form*. For example, they will give the *magnitude* of forward transadmittance as $|y_{fs}|$ and the *angle* of forward transadmittance of $/y_{fs}$. Quite often, manufacturers will mix the two systems of vector algebra on their datasheets.

Conversion of vector algebra forms. It is assumed that the readers are already familiar with the basics of vector algebra. However, the following notes summarize the steps necessary to manipulate vector algebra terms. With this background the reader should be able to perform all the calculations involved in the design of RF amplifier networks.

Converting from rectangular to polar form:
(1) Find the magnitude from the square root of the sum of the squares of the components.

$$\text{Polar magnitude} = \sqrt{g^2 + jb^2}$$

(2) Find the angle from the ratio of the component values.

$$\text{Polar angle} = \arctan \frac{jb}{g}$$

The angle is leading if the *jb* term is positive, and lagging if the *jb* term is negative.

For example, assume that the y_{fs} is given as $g_{fs} = 30$ and $jb_{fs} = 70$. This is converted to polar form by

$$|y_{fs}| \text{ polar magnitude} = \sqrt{(30)^2 + (70)^2} = 76$$

$$\underline{/y_{fs}} \text{ polar angle} = \arctan \frac{70}{30} = 67°$$

Converting from polar to rectangular form:
(1) Find the real (conductive or g) part when polar magnitude is multiplied by the cosine of the polar angle.
(2) Find the imaginary (susceptance or jb) part when polar magnitude is multiplied by the sine of the polar angle.

If the angle is positive, the jb component is also positive. When the angle is negative, the jb component is also negative.

For example, assume that the y_{fs} is given as $|y_{fs}| = 20$ and $\underline{/y_{fs}} = -33°$. This is converted to rectangular form by

$$20 \times \cos 33° = g_{fs} = 16.8$$
$$20 \times \sin 33° = jb_{fs} = 11$$

The four basic y-parameters. A y-equivalent circuit is shown in Fig. 2-4. This equivalent circuit is for an FET. However, a similar circuit can be drawn for a two-junction transistor when analyzing small-signal characteristics.

Note that y-parameters can be expressed with number subscripts or letter subscripts. The number subscripts are universal because they can apply to two-junction transistors, FETs, and even integrated circuit amplifiers. However, the letter subscripts are most popular on FET datasheets.

The following notes can be used to standardize y-parameter nomenclature

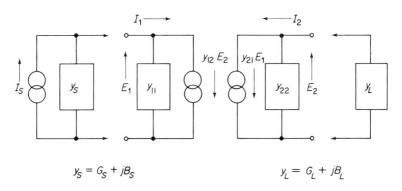

Figure 2-4 y-equivalent circuit (for an FET) with source and load

(note that the letter s in the letter subscript refers to common-source operation of an FET amplifier).

y_{11} is input admittance, and can be expressed as y_{is} on an FET datasheet.

y_{12} is reverse transadmittance, and can be expressed as y_{rs} on an FET datasheet.

y_{21} is forward transadmittance, and can be expressed as y_{fs} on an FET datasheet.

y_{22} is output admittance, and can be expressed as y_{os} on an FET datasheet.

Input admittance, with $Y_L = $ infinity (short circuit), is expressed as

$$y_{11} = g_{11} + jb_{11} = \frac{di_1}{de_1} \quad \text{(with } e_2 = 0\text{)}$$

This means that y_{11} is equal to the difference in current i_1, divided by the difference in voltage e_1, with voltage e_2 at 0. The voltages and currents involved are shown in Fig. 2-4.

Some datasheets do not show y_{11} at any frequency, but give input capacitance instead. If one assumes that the input admittance is entirely (or mostly) capacitive, then the input impedance can be found when input capacitance is multiplied by 6.28F ($F = $ frequency in hertz) and the reciprocal is taken. Because admittance is the reciprocal of impedance, admittance is found when input capacitance is multiplied by 6.28F (where admittance is capacitive). For example, if the frequency is 100 MHz and the input capacitance is 8 pF, the input admittance is

$$6.28 \times (100 \times 10^6) \times (8 \times 10^{-12}) \approx 5 \text{ mmhos}$$

This assumption is accurate only if the real part of y_{11} (or g_{11}) is negligible. Such an assumption is reasonable for an FET, but not necessarily for a two-junction transistor. Figure 2-5 shows input admittance curves for a typical FET. Note that the imaginary part (jb_{11}) is greater in value across the entire frequency range; however, the real part of two-junction transistor input admittance can be quite large in relation to the imaginary jb_{11} part.

Forward transadmittance, with $Y_L = $ infinity (short circuit), is expressed as

$$y_{21} = g_{21} + jb_{21} = \frac{di_2}{de_1} \quad \text{(with } e_2 = 0\text{)}$$

This means that y_{21} is equal to the difference in output current i_2, divided

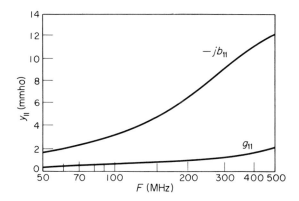

Figure 2-5 Input admittance $y_{11} = g_{11} + jb_{11}$ for typical FET

by the difference in input voltage e_1, with voltage e_2 at 0. In other words, y_{21} represents the difference in output current for a difference in input voltage.

Some FET datasheets show y_{21} at a low frequency (typically 1 kHz), and then show $R_e(y_{fs})$, or the real part of y_{21}, at a high frequency (typically 100 to 200 MHz). Other FET datasheets specify that g_{21} is the real part of y_{21} and that the values given are for a low frequency. Then some value is given for y_{21} at a high frequency.

Two-junction transistor datasheets often do not give any value for y_{21}. Instead, they show forward transadmittance by means of the hybrid system of notation using h_{fe} or h_{21} (which means hybrid forward transadmittance with common emitter). No matter what system is used, it is essential that the values of forward transadmittance be considered at the frequency of interest.

These considerations are illustrated in Fig. 2-6 which shows more accurate and complete forward transadmittance curves for a typical FET. Note from

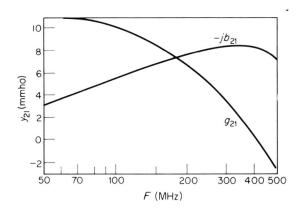

Figure 2-6 Forward transadmittance, $y_{21} = g_{21} + jb_{21}$ for typical FET

the figure that the real and imaginary parts intersect around 180 MHz, and that the real part becomes a negative quantity around 400 MHz.

Output admittance, with $Y_S = $ infinity (short circuit), is expressed as

$$y_{22} = g_{22} + jb_{22} = \frac{di_2}{de_2} \quad \text{(with } e_1 = 0\text{)}$$

Figure 2-7 shows output admittance curves for a typical FET. Note that the real part is negligible over the entire frequency range. This condition is not true for a typical two-junction transistor.

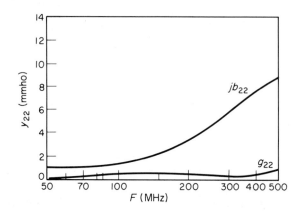

Figure 2-7 Output admittance $y_{22} = g_{22} = jb_{22}$ for typical FET

Reverse transadmittance, with $Y_S = $ infinity (short circuit), is expressed as

$$y_{12} = g_{12} + jb_{12} = \frac{di_1}{de_2} \quad \text{(with } e_i = 0\text{)}$$

y_{12} is not considered an important two-junction transistor parameter. Many FET datasheets do not list y_{12} at any frequency and give reverse transfer capacitance instead. If we assume that the reverse transadmittance (also known as *reverse transfer admittance*) is entirely (or mostly) capacitive (jb_{12}), then the reverse transadmittance can be found when reverse transfer capacitance is multiplied by $6.28F$ ($F = $ frequency in hertz). This assumption is generally accurate in the case of y_{12} for FETs, as shown in Fig. 2-8. Note that the real part of y_{12} (or g_{12}) is 0 across the entire frequency range. Thus, when the term $R_e(y_{12})$ appears in an equation (as it does frequently in FET RF design equations), the term can be considered as zero for all FET applications. This is not necessarily true for two-junction transistors.

y-parameter measurement. It is obvious that *y*-parameter information is not always available or in a convenient form. In practical design, it may be necessary to measure the *y*-parameters, using laboratory equipment.

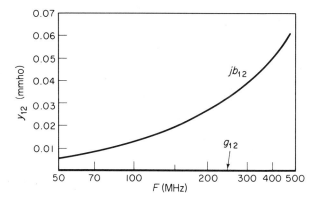

Figure 2-8 Reverse transfer admittance $y_{12} = g_{12} + jb_{12}$ for typical FET

Note that all y-parameters are based on *ratios* of input-output current to input-output voltage. For example, y_{21} is the ratio of output current to input voltage.

For FETs, y_{21} and y_{22} can be measured by using signal generators, voltmeters, and simple circuits. Likewise, y_{11} and y_{12} can be found by measuring c_{iss} and c_{rss} (input and reverse capacitances) using a simple capacitance meter, and then calculating y_{11} and y_{12} based on the frequency of interest. These procedures are described in the author's *Handbook for Transistors* (Prentice-Hall, Inc., Englewood Cliffs, N.J., 1976).

More accurate results can be obtained with precision laboratory equipment. All four y-parameters can be measured on a General Radio Transfer Function and Immittance Bridge. A possible exception is y_{12}, which is typically very much smaller than the other FET parameters. In the case of y_{12} it is often more practical to measure c_{rss} and multiply by 6.28F.

The main concern in measuring y-parameters, with FETs or two-junction transistors, is that the measurements are made under *conditions simulating those of the final circuit*. For example, if supply voltages, bias voltages, and operating frequency are not identical (or close) to the final circuit, the tests may be misleading.

Two-junction transistor admittance measurements. Although the datasheets for transistors to be used as power amplifiers usually contain input and output admittance information, it may be helpful to know how this information is obtained.

A typical test-amplifier circuit for two-junction transistors is shown in Fig. 2-9. During a test, the transistor is placed in the test circuit designed with *variable components* to provide wide tuning capabilities. This feature is necessary to insure correct matching while characterizing a transistor at serveral power levels. The circuit is tuned for maximum power gain at each power level for which admittance data is desired.

After the test amplifier has been tuned for maximum power gain, the d-c power, signal source, circuit load, and test transistor are disconnected from the circuit. For total circuit impedance to remain the same, the signal source and output load circuit connections are terminated at their characteristic resistances. After these substitutions are performed, complex admittances are measured at the base and collector circuit connections of the test transistor (points A and B, respectively, in Fig. 2-9).

The two-junction transistor input and output admittances are the *conjugates* of the base circuit connection and the collector circuit connection admittances, respectively. For example, if the base circuit connection admittance is $7 + j3$, the input admittance of the transistor is $7 - j3$.

In some systems of two-junction transistor RF amplifier design, the networks are calculated on the basis of input-output resistance and capaci-

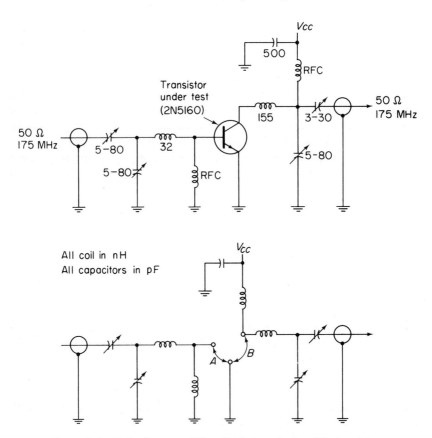

Figure 2-9 Typical test-amplifier circuit for two-junction transistors

tance, instead of admittance. In such cases, the admittances measured in the circuit in Fig. 2-9 must be converted to resistance and capacitance.

Admittances are expressed in mhos (or millimhos, mmhos). Resistance can be found by dividing the real part of the admittance into 1. Capacitance can be found by dividing the imaginary part of the admittance into 1 (to find reactance); then the reactance is used in the equation $C = 1/(6.28F X_C)$ to find actual capacitance.

2-2.2. Stability Factors

There are two factors used to determine the potential stability (or instability) of transistors in RF amplifiers. One factor is known as the Linvill C factor; the other is the Stern k factor. Both factors are calculated from equations requiring y-parameter information (to be taken from datasheets or by actual measurement at the frequency of interest).

The main difference between the two factors is that the Linvill C factor assumes the transistor is not connected to a load. The Stern k factor includes the effect of a specific load.

The Linvill C factor is calculated from

$$C = \frac{y_{12}y_{21}}{2g_{11}g_{22} - R_e(y_{12}y_{21})}$$

If C is less than 1, the transistor is unconditionally stable. That is, using a conventional (unmodified) circuit, no combination of load and source admittance can be found which will cause oscillation. If C is greater than 1, the transistor is potentially unstable. That is, certain combinations of load and source admittance will cause oscillation.

The Stern k factor is calculated from

$$k = \frac{2(g_{11} + G_S)(g_{22} + G_L)}{y_{12}y_{21} + R_e(y_{12}y_{21})}$$

where G_S and G_L are source and load conductance, respectively. (G_S = 1/source resistance; G_L = 1/load resistance.)

If k is greater than 1, the amplifier circuit is stable (the opposite of Linvill). If k is less than 1, the amplifier is unstable. In practical design, it is recommended that a k factor of 3 or 4 be used, rather than 1, to provide a margin of safety. This will accommodate parameter and component variations (particularly with regards to band-pass response).

Note that both equations are fairly complex, and require considerable time for their solutions. In practical work, computer-aided design techniques are used for stability equations.

Some manufacturers provide alternative solutions to the stability and

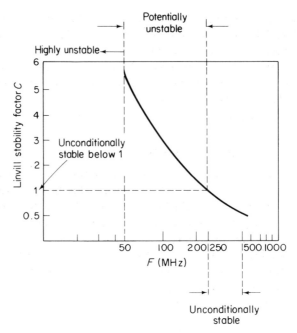

Figure 2-10 Linvill stability factor C, for typical FET

load-matching problem, usually in the form of a datasheet graph, as shown in Fig. 2-10. This is a Linvill C factor chart for a typical MOSFET. Note that the MOSFET is unconditionally stable at frequencies above 250 MHz. At frequencies below about 50 MHz, the MOSFET becomes highly unstable.

2-2.3. Solutions to Stability Problems

There are two basic design solutions to the problem of unstable RF amplifiers.

First, the amplifier can be neutralized; that is, part of the output can be fed back (after it is shifted in phase) to the input so as to cancel oscillation. Neutralization permits the amplifier to be matched perfectly to the source and load. This type of match is known as a *conjugate match*. In perfect conjugate match the transistor input and source, as well as the transistor output and load, are *matched* resistively, and *all reactance* is tuned out. Neutralization requires extra components and creates a problem when frequency is changed.

The second solution is to *introduce some mismatch* into either the source or load tuning networks. This solution, sometimes known as the *Stern solution*, requires no extra components, but does produce a reduction in gain.

A comparison of these two methods is shown in Fig. 2-11. The higher gain curve represents the unilateralized (or neutralized) operation. The lower gain curve represents the circuit power gain when the Stern k factor is 3.

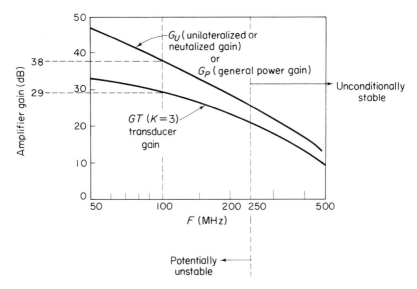

Figure 2-11 Comparison of neutralized gain (G_U) versus gain with mismatch (G_T) when the Stern k factor is 3

Assume that the frequency of interest is 100 MHz. If the amplifier is matched directly to the load (perfect conjugate match) without regard to stability or using neutralization to produce stability, the top curve applies and the power gain is about 38 dB. If the amplifier is matched to a load and source where the Stern k factor is 3 (resulting in a mismatch with the actual load and source), the lower curve applies and the power gain is about 29 dB.

The upper curve of Fig. 2-11 is found by the *general power gain equation*

$$G_P = \frac{\text{power delivered to load}}{\text{power delivered to input}}$$

$$= \frac{(y_{21})^2 G_L}{(Y_L + y_{22})^2 R_e\left(y_{11} - \dfrac{y_{12} y_{21}}{y_{22} + Y_L}\right)}$$

The general power gain equation applies to circuits with no external feedback, and to circuits which have external feedback (neutralization)—provided the composite y-parameters of both transistor and feedback networks are substituted for the transistor y-parameters in the equation.

The lower curve in Fig. 2-11 is found by the *transducer gain equation*

$$G_T = \frac{\text{power delivered to load}}{\text{maximum power available from source}}$$

$$= \frac{4 G_s G_L (y_{21})^2}{[(y_{11} + Y_s)(y_{22} + Y_L) - y_{12} y_{21}]^2}$$

The transducer gain expression includes input mismatch. The lower curve in Fig. 2-11 assumes that the input mismatch produces a Stern k factor of 3. That is, the circuit tuning networks are adjusted for admittances that produce a Stern k factor of 3. The transducer gain expression considers the input and output networks as part of the source and load.

With either gain expression, the input and output admittances of the transducer are modified by the load and source admittances.

The input admittance of the transistor is given by

$$Y_{IN} = y_{11} - \frac{y_{12} y_{21}}{y_{22} + Y_L}$$

The output admittance of the transistor is given by

$$Y_{OUT} = y_{22} - \frac{y_{12} y_{21}}{y_{11} + Y_S}$$

At low frequencies, the second term in the input and output admittance equations is not particularly significant. At VHF, the second term makes a very significant contribution to the input and output admittances.

The imaginary parts of Y_S and Y_L (B_S and B_L, respectively) must be known before values can be calculated for power gain, transducer gain, input admittance, and output admittance. Exact solutions for B_S and B_L almost always consist of time consuming complex algebraic manipulations.

To find fairly good simplifying approximations for the equations, let $B_S \approx -b_{11}$ and $B_L \approx -b_{22}$ so that

General power gain expression

$$G_P \approx \frac{(y_{21})^2 G_L}{(G_L + g_{22})^2 R_e \left(y_{11} - \frac{y_{12} y_{21}}{g_{22} + G_L} \right)}$$

Transducer gain expression

$$G_T \approx \frac{4 G_S G_L (y_{21})^2}{[(g_{11} + G_S)(g_{22} + G_L) - y_{12} y_{21}]^2}$$

Input admittance

$$Y_{IN} \approx y_{11} - \frac{y_{12} y_{21}}{g_{22} + G_L}$$

Output admittance

$$Y_{OUT} \approx y_{22} - \frac{y_{12} y_{21}}{g_{11} + G_S}$$

The other gain expressions often found on the datasheets of transistors used in RF applications (particularly FET datasheets) are: maximum available gain (MAG), and maximum usable gain (MUG).

MAG is usually applied as the gain in a conjugately matched, neutralized circuit and is expressed as

$$\text{MAG} = \frac{(y_{21})^2 R_{IN} R_{OUT}}{4}$$

where R_{IN} and R_{OUT} are the input and output resistances, respectively, of the transistor.

An alternate MAG expression is

$$\text{MAG} = \frac{(y_{21})^2}{4R_e(y_{11})R_e(y_{22})}$$

where $R_e(y_{11})$ is the real part (g_{11}) of the input admittance, and $R_e(y_{22})$ is the real part (g_{22}) of the output admittance.

MUG is usually applied as the *stable gain* which may be realized in a *practical* (neutralized or unneutralized) RF amplifier. In a typical unneutralized FET circuit, MUG is expressed as:

$$\text{MUG} \approx \frac{0.4 y_{21}}{6.28 F\, c_{rss}}$$

where c_{rss} is reverse transfer capacitance.

MAG and MUG are often omitted on datasheets for two-junction transistors; instead, gain is listed as h_{fe} at a given frequency. This is supplemented with graphs which show available power output at given frequencies with a given input.

2-2.4. Neutralized Solution

There are several methods for neutralization of RF amplifiers. The most common method is the *capacitance-bridge* technique, as shown in Fig. 2-12(a). Capacitance-bridge neutralization becomes more apparent when the circuit is redrawn, as shown in Fig. 2-12(b). The condition for neutralization is that $I_F = I_N$.

The equations normally used to find the value of the feedback neutralization capacitor are long and complex. However, for practical work, if the value of C_1 is made quite large in relation to C_2 (at least four times), the value of C_N can be found by

$$C_N \approx C_F \frac{C_1}{C_2}$$

where C_F is the reverse capacitance of the transistor.

In simple terms, the value of C_N is approximately equal to the value of reverse capacitance times the ratio C_1/C_2. For example, if reverse capacitance

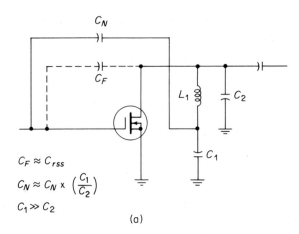

$C_F \approx C_{rss}$

$C_N \approx C_N \times \left(\dfrac{C_1}{C_2}\right)$

$C_1 \gg C_2$

(a)

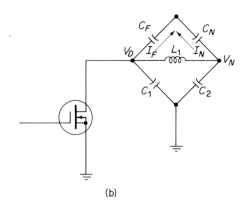

(b)

Figure 2-12 Capacitance-bridge neutralization circuit

is 0.2 pF, C_1 is 30 pF and C_2 is 3 pF; the C_1/C_2 ratio is 10, and $C_N \approx 10 \times 0.2 \text{ pF} = 2 \text{ pF}$.

2-2.5. The Stern Solution

A stable design with a potentially unstable transistor is possible without external feedback (neutralization) by proper choice of source and load admittances. This can be seen by inspection of the Stern k factor equation; G_S and G_L can be made large enough to yield a stable circuit, regardless of the degree of potential instability. Using this approach, a circuit stability factor k (typically $k = 3$) is selected, and the Stern k factor equation is used to arrive at values of G_S and G_L which will provide the desired k.

Of course, the actual G of the source and load cannot be changed; instead, the input and output tuning circuits are designed as if the actual G values were changed. This results in a mismatch and a reduction in power gain, but does produce the desired degree of stability.

To get a particular circuit stability factor, the designer may choose any of the following combinations of matching or mismatching of G_S and G_L to the transistor input and output conductances, respectively:

G_S matched and G_L mismatched
G_L matched and G_S mismatched
Both G_S and G_L mismatched

Other performance requirements or practical considerations often dictate the decision on which combination to use. For example, it may not be practical to mismatch to some extreme value of G_S or G_L.

Once G_S and G_L have been chosen, the remainder of the design may be completed by using the relationships that apply to the amplifier without feedback. Power gain and input-output admittances may be computed using the appropriate equations (Sec. 2-2.3).

Simplified Stern approach. Although the above procedure may be adequate in many cases, a more systematic method of source and load admittance determination is desirable for designs which demand maximum power gain per degree of circuit stability. Stern has analyzed this problem and developed equations for computing the best G_S, G_L, B_S and B_L for a particular circuit stability factor (Stern k factor). Unfortunately, these equations are very complex and become quite tedious when they are used frequently. The complete Stern solution is best applied by computer.

Programs have been written to provide essential information for transistors used as RF amplifiers, including the effects of various specific sources and loads. These programs permit the designer to experiment with theoretical "breadboard" circuits in a matter of seconds.

When a Stern solution must be obtained *without the aid of a computer*, it is best to use one of the many shortcuts that have been developed over the years. The following shortcut is by far the simplest and most widely accepted, yet provides an accuracy close to that of the computer solutions.

(1) Let $B_S \approx -b_{11}$ and $B_L \approx b_{22}$, as in the case of the Sec. 2-2.3 equations. This method permits the designer to closely approximate the exact Stern solution for Y_S and Y_L, while avoiding that portion of the computations which is the most complex and time consuming. Further, the circuit can be designed with tuning adjustments for varying B_S and B_L, thereby creating the possibility of achieving the true B_S and B_L (by experiment) for maximum gain as accurately as if all the Stern equations had been solved.

(2) Mismatch G_S to g_{11} and G_L to g_{22} by an *equal ratio*. That is, find a ratio that produces the desired Stern k factor, then mismatch G_S to g_{11} (and

G_L to g_{22}). For example, if the ratio is 4 to 1, make G_S 4 times the value of g_{11} (and G_L 4 times the value of g_{22}).

If the mismatch ratio, R, is defined as

$$R = \frac{G_L}{g_{22}} = \frac{G_S}{g_{11}}$$

then R may be computed for any particular circuit stability k factor using the equation

$$R = \left(\sqrt{k\left[\frac{y_{21}y_{12} + R_e(y_{12}y_{21})}{2g_{11}g_{22}}\right]}\right) - 1$$

As an example, assume that it is desired to mismatch input and output circuits so there is a Stern k factor of 4, using a transistor that has the following characteristics: $y_{12}y_{21} = 0.5$, $g_{11} = 5.0$, $g_{22} = 0.05$, $R_e(y_{12}y_{21}) = 0.2$ (all values in mmhos).

$$R = \left(\sqrt{4\left[\frac{0.5 + 0.2}{2(5)(0.05)}\right]}\right) - 1 \approx 1.37$$

Using the value of 1.37 for R, and the equation

$$1.37 = \frac{G_S}{g_{11}} = \frac{G_L}{g_{22}}$$

then

$$G_S = (1.37)(5)(10^{-3}) = 6.85 \text{ mmhos}$$

and

$$R_S = \frac{1}{G_S} \approx 146 \, \Omega$$

$$G_L = (1.37)(0.05)(10^{-3}) = 0.0685 \text{ mmho}$$

and

$$R_L = \frac{1}{G_L} \approx 14{,}600 \, \Omega$$

The shortcut Stern method may be advantageous if the source and load admittances and power gains for several different values of k are desired. Once the R for a particular k has been determined, the R for any other k may be quickly found from the equation

$$R = \frac{(1 + R_1)^2}{(1 + R_2)^2} = \frac{k_1}{k_2}$$

where R_1 and R_2 are values of R corresponding to k_1 and k_2, respectively.

Sec. 2-2 Basic RF Amplifier Design Approaches 47

The Stern solution with datasheet graphs. It is obvious that the Stern solution, even with the shortcut method, is somewhat complex. For this reason, some manufacturers have produced datasheet graphs that show the best source and load admittances for a particular transistor over a wide range of frequencies. Examples of these graphs are shown in Figs. 2-13 and 2-14.

Figure 2-13 shows both the real (G_S) and imaginary (B_S) values that will produce maximum gain, but with a stability (Stern k) factor of 3 at frequencies from 50 to 500 MHz. Figure 2-14 shows corresponding information for G_L and B_L.

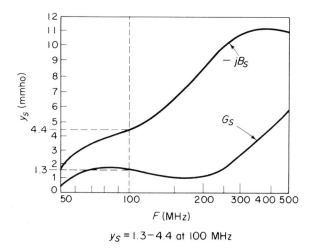

$y_S = 1.3 - 4.4$ at 100 MHz

Figure 2-13 Best source admittance, $Y_S = G_S + jB_S$

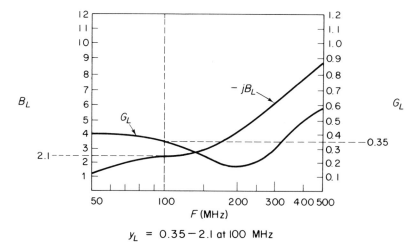

$y_L = 0.35 - 2.1$ at 100 MHz

Figure 2-14 Best load admittance, $Y_L = G_L + jB_L$

To use these figures, simply select the desired frequency, and note where the corresponding G and B curves cross the frequency line. For example, assuming a frequency of 100 MHz, $Y_L = 0.35 - j2.1$ mmhos, and $Y_S = 1.3 - j4.4$ mmhos.

If the tuning circuits are designed to match these admittances, rather than the actual admittances of the source and load, the circuit will be stable. Of course, the gain will be reduced. Use the *transducer gain expression* (G_T) of Sec. 2-2.3 to find the resultant power gain.

2-3. TYPES OF RF AMPLIFIER TUNING NETWORKS

As far as the tuning networks are concerned, there are only three basic types of RF amplifiers. These include voltage amplifiers, power amplifiers, and multipliers. Going further, there are only two basic methods by which the tuning networks match transistor impedances to source and load impedances. One method uses combinations of coils and capacitors which are tuned to cancel out impedance differences. The other method uses taps on the tank coils to match impedances. The basic theory for the various types is discussed in this section. In Chapter 6, the design requirements and specific design examples for the basic types are covered.

2-3.1. RF *Voltage Amplifier*

The circuit of Fig. 2-15 is a typical narrowband RF amplifier, such as those found in broadcast- and communications-type radio receivers. The circuit is a single stage of tuned RF voltage amplification. Input to the transistor is by means of a tuned RF transformer, and output is obtained by a similar device. Transformer T_1 is the input transformer, and T_2 is the output transformer. The secondary winding of T_1 is tuned to resonance at the frequency of the incoming signal by means of variable capacitor C_1. The primary of T_2 is tuned to the same resonant frequency by means of variable capacitor C_2.

At the resonant frequency, the secondary of T_1 and capacitor C_1 form a parallel-resonant circuit, as do the primary of T_2 and capacitor C_2. A parallel-resonant circuit offers a very high impedance to a current at the resonant frequency, but a low impedance at other frequencies. Thus, if C_1 is adjusted to tune the secondary winding of T_1 to resonance at the frequency of the desired signal, a relatively large voltage will appear across this resonant circuit (and the transistor base) for signals of this frequency. For signals of other frequencies, the voltage will be low.

Likewise, if C_2 is adjusted to tune the primary winding of T_2 to resonance at the frequency of the desired signal, this resonant circuit will show a large

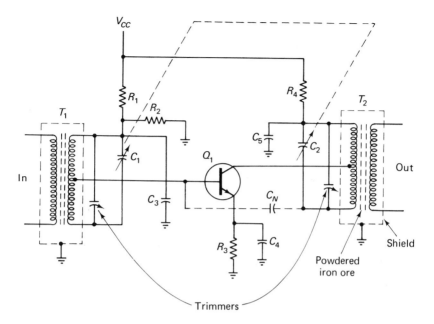

Figure 2-15 Tuned RF voltage amplifier

impedance for signals of this frequency, and a very low impedance for other signals. This resonant circuit is the *collector load*. As discussed in other chapters, voltage amplification of a stage is set by collector load impedances (all other factors remaining equal). These conditions provide high voltage amplification at the desired frequency. At all other frequencies, where the collector load impedance is low, the amplification will also be low.

The shunting effect of the transistor input and output capacitances is also minimized by the capacitances of the tuned resonant circuits. For example, the typically small input capacitance of the transistor is in parallel with the relatively large capacitance of variable capacitor C_1, and thus has but a small additive effect.

Bias networks. Resistors R_1 and R_2 form a voltage divider across the power supply (V_{CC}) to bias the emitter-base junction. The bias circuits for RF amplifiers are essentially the same as for audio amplifiers described in Chapter 5. However, the operating point may be different. Typically, audio amplifiers are operated as class A, where collector current flows at all times. RF amplifiers are generally operated as class B or C, where current flows only in the presence of a signal.

Resistor R_2 and capacitor C_3 form a decoupling network to prevent the RF signal from entering the power supply (through which the signal may be fed to the output circuit, or another stage). Resistor R_3 is the emitter stabilization resistor and C_4 is its bypass capacitor. Resistor R_4 and capacitor C_5 form the decoupling network for the collector circuit.

Impedance match. Note that the base of the transistor is connected to a tap on the secondary winding of T_1. Since the input impedance of the transistor is relatively small, only a portion of the secondary winding is used to obtain a proper impedance match between the two. However, the entire secondary winding is tuned by C_1 to form the parallel-resonant circuit at the signal frequency. For similar reasons the collector is connected to a tap on the primary winding of T_2.

Feedback problems. One difficulty often found in RF amplifiers is the prevention of feedback from the output of a stage to its own input, or to another stage. There are two types of undesired feedback: *radiated feedback* and *feedback through the transistor*. (Of course, there is feedback that is deliberately introduced to stabilize gain, temperature response, etc.)

To eliminate radiated feedback, the amplifier must be properly shielded and designed so as to separate the base and collector leads. The problem of RF amplifier shielding is extensive and will not be covered here because we are primarily concerned with transistors.

Fortunately, most modern transistors are so constructed that there is little danger of feedback through the transistor at moderately high frequencies. However, as frequency increases, internal feedback can produce undesired conditions in an RF amplifier.

One feedback problem is known as the *Miller effect*. As shown in Fig. 2-16, there is a capacitance between base and emitter of a two-junction transistor (or between gate and source of an FET). This forms the input capacitance of the circuit. There is also a capacitance between the base and collector (or gate and drain). This capacitance feeds back some of the collector signal to the base. The collector signal is amplified, and is 180 degrees out of phase with the base signal (in a common-emitter amplifier). The collector signal feedback opposes the base signal and tends to distort the input signal. Likewise, the collector-base capacitance is, in effect, in series with the base-emitter capacitance, and thus changes the input capacitance.

These conditions result in a constantly changing amplitude-modulated relationship of signals in an amplifier. For example, if the input signal amplitude changes, the amount of feedback changes, changing the input capacitance. In turn, the change in input capacitance changes the match between the transistor and the input tuned circuit, changing the amplitude. Likewise, if the input signal frequency changes, the feedback changes (since the collector-base capacitive reactance changes), and there is a corresponding change in amplification.

The Miller effect is not necessarily a problem in all solid-state RF amplifiers. The FET RF amplifier is usually more susceptible to Miller effect than

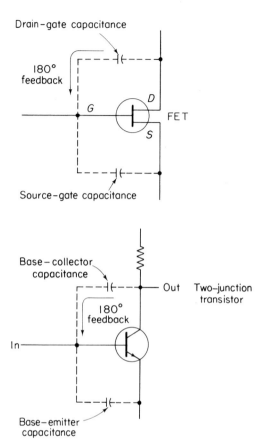

Figure 2-16 Input and feedback capacitances in transistors

two-junction transistors. However, when the Miller effect becomes severe with any RF amplifier, it can be eliminated or minimized to a realistic level by *neutralization*.

Neutralization is a method for reducing the amount of unwanted feedback, either from radiation or internal feedback. With neutralization, a portion of the voltage from the output circuit of the stage is fed back to the input circuit in such a way that it cancels the base voltage caused by the unwanted feedback. Neutralization is accomplished by impressing a voltage on the base that is equal in magnitude, but opposite in phase, to the undesired feedback. Thus, the two voltages will "buck" each other out.

The two ends of the primary winding of the output transformer (such as T_2 in Fig. 2-15) are of opposite phase. If this opposite phase voltage is fed to the base through the neutralizing capacitor (C_N of Fig. 2-15), the two voltages will cancel out.

Another method for reducing the unwanted feedback, without neutraliza-

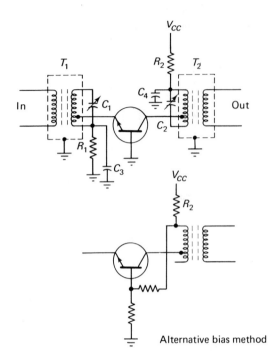

Figure 2-17 Common-base (grounded-base) RF amplifier
Alternative bias method

tion, is to use the common-base (or common-gate) amplifier configuration. A common-base RF amplifier circuit is shown in Fig. 2-17. The base is grounded. The input signal is applied to the emitter. The output is obtained between the collector and base, which is common to the input and output circuits. The grounded base acts as a shield between the input and output circuits, thus reducing feedback.

2-3.2. RF *Power Amplifier*

Most radio transmitters use some form of power amplifier to raise the low-amplitude signal developed by the oscillator to a high-amplitude signal suitable for transmission. For example, most oscillators develop signals of less than 1 W, whereas a solid-state transmitter may require 100-W (or more) output.

Figure 2-18 shows two basic RF power-amplifier circuits. In the circuit of Fig. 2-18(a), the collector load is a parallel-resonant circuit (called a *tank circuit*) consisting of variable capacitor C_1 and inductor L_1, tuned to resonance at the desired frequency. The output, which is an amplified version of the input voltage, is from L_2, which together with L_1 forms an output transformer.

The circuit of Fig. 2-18(a) has both advantages and disadvantages. The

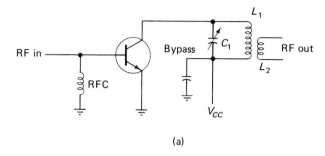

(a)

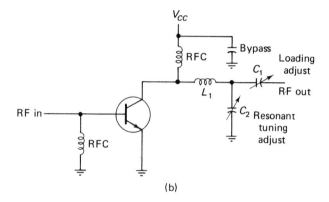

(b)

Figure 2-18   Radio-frequency power amplifiers

winding of L_2 can be made to match the impedance of the load (by selecting the proper number of turns and by positioning L_2 in relation to L_1). While that may prove an advantage in some cases, it also makes for an interstage coupling network that is subject to mismatch and detuning by physical movement of shock.

Another disadvantage of the Fig. 2-18(a) circuit is that all the current must pass through the tank circuit coil. Also, for best transfer of power the impedance of L_1 should match that of the transistor output. Since two-junction transistor output impedances are generally low, the value of L_1 must be low, often resulting in an impractical size for L_1. The circuit of Fig. 2-18(a) is a carry-over from vacuum tube circuits and, as such, is not often found in modern two-junction transistor amplifiers. However, the circuit is found in FET RF amplifiers (which are generally low power and higher impedance).

The circuit of Fig. 2-18(b), or one of its many variations, is commonly found in solid-state radio transmitters using two-junction transistors. The collector load is a resonant circuit formed by the network of L_1, C_1 and C_2.

Note that C_1 is marked "Loading adjust," whereas C_2 is marked "Resonant tuning adjust." These networks provide the dual function of frequency selection (equivalent to the tank circuit) and impedance matching between transistor and load. For impedances to be properly matched, both the resistive (so-called real part) and reactive (so-called imaginary part) components of the impedance must be considered.

Both Fig. 2-18 circuits are operated class B, which is typical for RF amplifiers. Class B operation is obtained by connecting the emitter directly to ground and applying no bias to the base-emitter junction. Since any two-junction transistor requires some forward bias to produce current flow, the transistor remains cut off except in the presence of a signal.

2-3.3. RF *Multiplier*

The circuits of Fig. 2-18 can be used as a frequency multiplier. That is, the collector circuit is tuned to a higher whole-number multiple (harmonic) of the input frequency. Many radio transmitters use some form of multiplier to raise the low-frequency signal developed by the oscillator to a high-frequency signal. For example, most crystals used in oscillators have a fundamental frequency of less than 10 MHz, whereas a solid-state transmitter may produce an output in the UHF range.

Although the circuits of power multipliers and power amplifiers are essentially the same, the efficiency is different. An amplifier operating at the same frequency as the input signal will have a higher efficiency than an identical circuit operating at a multiple of the input frequency.

2-3.4. RF *Amplifier-Multiplier Combinations*

The circuits of Fig. 2-18 can be cascaded to provide increased power amplification and/or frequency multiplication. Typically, no more than three stages are so cascaded. The stages can be mixed. That is, one or two stages can provide frequency multiplication, with the remaining one or two stages providing power amplification.

2-4. DESIGN OF RF NETWORKS USING VOLTAGE VARIABLE CAPACITORS

Voltage variable capacitors (VVC) are used in many applications to tune RF networks. The capacitance of a VVC is controlled by an external voltage. That is, the capacitance can be varied when the external voltage is varied. If a VVC is used in an RF tuning network, it is possible to vary the network capacitance, and thus vary the network resonant frequency, with a variable external voltage.

Sec. 2-4 Design of RF Networks Using Voltage Variable Capacitors

A VVC is sometimes called a *voltage variable diode* since the device is constructed more like a diode than a capacitor. However, VVC is the accepted term. The reader can obtain a more comprehensive discussion of the theory of VVC operation by consulting the author's *Practical Semiconductor Databook for Electronic Engineers and Technicians* (Prentice-Hall, Inc., Englewood Cliffs, N.J., 1970).

The main concern in designing with VVCs in any resonant circuit is the *tuning range* of the circuit. All other factors being equal, the tuning range is dependent on the capacitance range of the VVC. Motorola has developed a graph that shows how tuning range can be predicted using the VVC and external circuit parameters.

Most VVC resonant circuits are in the form shown in Fig. 2-19 for parallel circuits, and in the form shown in Fig. 2-20 for series circuits. The effective circuit inductance is given by L. In some cases, for biasing purposes, there are additional RF chokes which, if properly chosen, have little effect on the resonant frequency. Circuit capacity shunting the VVC is given by C_{ckt}. The VVC capacity is given by $C_c + C_j$, the sum of case and junction capacitances.

Note: It is assumed that the resonant frequencies are well below the VVC

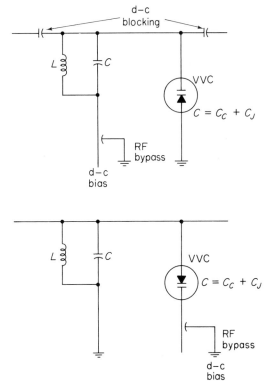

Figure 2-19 Typical parallel circuits for VVC control

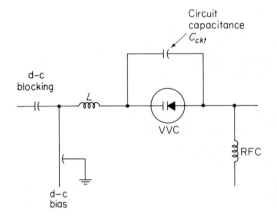

Figure 2-20 Typical series circuit for VVC control

self-resonant frequency so that any inductance can be ignored. It is not recommended that the VVC be operated in a circuit near the self-resonant frequency.

One of the steps in finding the tuning range of a VVC resonant circuit is to find the VVC capacity at various voltages. Most VVC specification sheets list the capacity at one voltage only. The capacity at other voltages can be calculated using the ratio of the two voltages. The basic equation is

$$\text{Voltage ratio} = \sqrt{\frac{1 + \frac{\text{known voltage}}{0.5}}{1 + \frac{\text{unknown voltage}}{0.5}}}$$

where *known voltage* is the voltage where the capacity is known, and *unknown voltage* is the voltage where the capacity is not known.

The basic equation also assumes a contact potential of 0.5 V, and a 0.5 power law of the VVC junction (typical for silicon VVCs).

For example, assume that it is desired to known the capacity of a VVC at -2 V, if the capacity is 22 pF at -4 V. Since -2 V is less reverse bias than -4 V, the capacity will be *increased* (by the ratio of the two voltages)

$$\sqrt{\frac{1 + \frac{4}{0.5}}{1 + \frac{2}{0.5}}} = 1.34 \text{ ratio}$$

$$1.34 \times 22 \text{ pF} = 29.48 \text{ pF}$$

VVC tuning range problems can be solved using the graph in Fig. 2-21. The information there can be used with any VVC, as long as the problem is solved in terms of minimum-to-maximum *capacity* ratio.

Sec. 2-4 Design of RF Networks Using Voltage Variable Capacitors 57

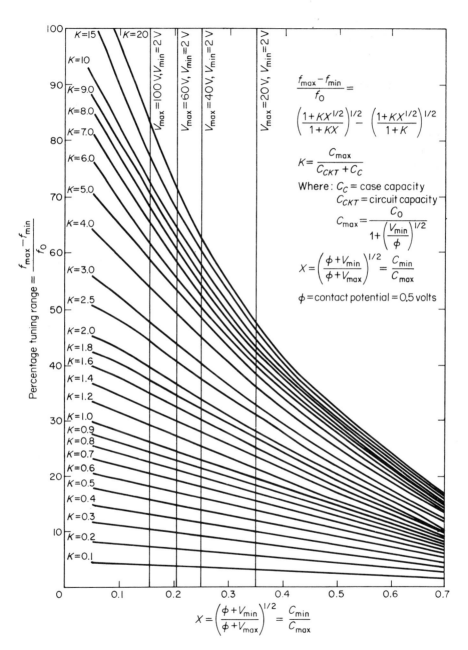

Figure 2-21 Percentage resonant-circuit tuning range for VVC (Courtesy Motorola)

2-4.1. Using the VVC Resonant Circuit Graph

The following is an example of how the graph in Fig. 2-21 can be used to find tuning characteristics of VVC circuits. Assume that the desired tuning range is between 60 and 90 MHz, and the circuit has a fixed value of 10 pF.

The equation of Fig. 2-21 shows that the percentage tuning range equals $(f_{max} - f_{min})/f_0$. f_{max} is 90 MHz, $f_{min} = 60$ MHz, and $f_0 = \sqrt{60 \times 90} \approx 73$ MHz. Thus, the percentage tuning range is $(90 - 60)/73 = 0.41$, or 41 per cent.

The maximum usable voltage (V_{max}) for a typical VVC is 60 V. That is, the reverse voltage must not be more negative than -60 V. Thus, the most negative reverse voltage determines the maximum usable voltage.

The opposite voltage limit (nearest to zero volts) is determined by temperature stability and/or intermodulation effects because the VVC junction capacity varies sharply at low voltages. Temperature dependence enters via the contact potential, which is more significant with low applied bias voltages. It is generally accepted that a lower limit (V_{min}) of -2 V is used for most applications.

As shown in Fig. 2-21, with a percentage tuning range of 41, a V_{max} of 60, and a V_{min} of 2 V, the nearest constant K is 2.5.

Using a fixed circuit value C_{ckt} of 10 pF, a C_c of 0.3 pF (which is typical for a glass VVC), and constant-K of 2.5, calculate C_{max} as follows

$$C_{max} = K(C_{ckt} + C_c) = 2.5(10 + 0.3) \approx 26 \text{ pF}$$

Note that this C_{max} value is for a voltage of -2 V. If the VVC datasheet lists another voltage for a given capacity, C_{max} must be related to that voltage. This can be done by calculating the ratio of the two voltages, as previously discussed, and then dividing the -2-V C_{max} by this ratio. For example, if the VVC datasheet shows a 22-pF capacity at -4 V, the ratio is 1.34. Using this ratio, and the -2-V C_{max} of 26 pF, the -4-V C_{max} is: 26 pF/1.34 = 19.4 pF for -4 V.

Any VVC having a -4-V capacity greater than 19.4 pF will tune the desired range. Allowing for a standard tolerance of ± 10 per cent in capacity, a VVC with a nominal -4-V capacity of 22 pF (19.8 to 24.2) will meet most requirements. Where circuit Q is important (Sec. 2-1.1), the lowest possible capacity should be used.

In some design situations, it is necessary to sacrifice minimum capacity (and best circuit Q). An example of this is where best temperature stability is desired. A VVC has best temperature stability where the minimum voltage is as high as possible. A similar case occurs when the maximum available control voltage is limited. Either situation increases the constant-K factor. For example, if the maximum available control voltage is 20 V, a K of about

Sec. 2-4 Design of RF Networks Using Voltage Variable Capacitors

6 is required, assuming the same percentage tuning range of 41. If K is increased, C_{max} is increased, and a higher value VVC must be used, thus reducing circuit Q.

The value of the coil to be used with the VVC can be determined using the equation

$$L \text{ (in } \mu\text{H)} = \frac{2.54 \times 10^4}{F \text{ (in MHz)}^2 \times C \text{ (in pF)}}$$

where L is the value of the coil, F is the *minimum* frequency, and C is the *total capacity* at the minimum voltage.

Using the previous example, the minimum frequency is 60 MHz, and the total capacitance at -2 V is 39.8 pF (22 pF × 1.34 ratio + 10-pF circuit capacity + 0.3-pF case capacity). Substituting these values, the value of L is

$$L \text{ (in } \mu\text{H)} = \frac{2.54 \times 10^4}{60^2 \times 39.8} \approx 0.18 \ \mu\text{H}$$

The maximum voltage required for the circuit can also be found by using the graph in Fig. 2-21. First find a new value of K using the 22-pF VVC (at -4 V) and the 1.34 ratio

$$K = \frac{22 \times 1.34}{10 + 0.3} = 2.85$$

Refer to Fig. 2-21 with a 41 per cent tuning range, and the new K of 2.85 to find an X (horizontal axis) of about 0.24.

Using the X of 0.24, the *minimum* allowable K of 2.5, and the contact potential of 0.5, find V_{max} as follows

$$V_{max} = \frac{\text{minimum allowable } K(2.5)}{X \text{ factor squared } [(0.24)^2]} - \text{contact potential } (0.5) = 43 \text{ V}$$

3 WAVE-GENERATING CIRCUIT DESIGN

Virtually all of the classic vacuum tube wave-generating circuits can be duplicated with transistors. This includes sawtooth oscillators, multivibrators, and Schmitt triggers. In many cases, transistor wave-generating circuits are superior to vacuum tube circuits. Likewise, certain types of transistors are particularly well suited to specific wave-generating circuits. For example, the UJT (unijunction transistor) is generally an excellent source for sawtooth waveforms. In this chapter, we shall discuss those transistor wave-generating circuits which have proven their value over the years.

3-1. SAWTOOTH OSCILLATORS

Both two-junction and FET transistors can be used in sawtooth oscillator circuits. However, the UJT has certain characteristics that make it well suited for use as a sawtooth waveform source. The *voltage waveform at the emitter* of the UJT in the basic relaxation oscillator is a fair approximation of a sawtooth waveform. However, if the emitter output of a UJT oscillator is connected directly to a load (either inductive or resistive), the circuit may fail to oscillate. Even if oscillation continues, the waveform will probably be distorted.

The most practical method to couple the emitter output of a UJT oscillator to a load is by means of a direct-coupled emitter follower. Such a circuit is shown in Fig. 3-1. Note that simple direct coupling is made possible by the fact that typical minimum voltage at a UJT emitter $V_{E(\min)}$ is about 1.2 V. If

Sec. 3-1   Sawtooth Oscillators

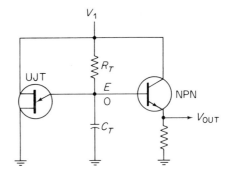

$$\frac{(\beta + 1)R_L}{R_T + (\beta + 1)R_L} > \text{maximum standoff ratio}$$

$R_T \approx$ load impedance

$R_T \approx 0.1 \text{ to } 0.2 \times (\beta + 1)R_L$

$$C_T \approx \frac{\text{period}}{R_T + \ln \frac{1}{1 - \text{standoff ratio}}}$$

$V_{OUT} \approx E_0 \times 0.9 \approx V_1 \times 0.3$

$$V_1 \approx \frac{V_{OUT} \text{(desired)}}{0.3}$$

Figure 3-1 Basic UJT sawtooth oscillator

$V_{E(\min)}$ is less than the normal base-to-emitter drop of the transistor, then the waveform across the load R_L will be clipped. However, a typical base-emitter drop of a silicon transistor is about 0.7 V (or considerably less than 1.2 V).

3-1.1. Design Considerations for UJT Sawtooth Oscillators

The primary consideration for the circuit in Fig. 3-1 is the loading effect of the emitter follower. A small amount of loading will shift oscillator frequency. A large amount of loading will stop oscillation. This can be shown by the equivalent circuit in Fig. 3-2.

The loading effect of the emitter-follower stage is approximated by an equivalent circuit $(\beta + 1)R_L$ across capacitor C_T, where β is the d-c common-emitter current gain of the transistor. It is seen from this equivalent circuit that loading will change the frequency of oscillation, since the capacitor charging circuit will be changed by the presence of the resistor $(\beta + 1)R_L$.

To minimize the effects of loading on the frequency, the values of R_L and β should be as large as possible. If the value of β or R_L is too small, the circuit will not oscillate. To ensure oscillation, β and R_L must satisfy the condition

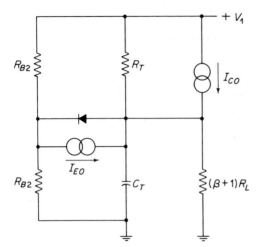

Figure 3-2   Equivalent circuit of UJT sawtooth oscillator

$$\frac{(\beta + 1)R_L}{R_T + (\beta + 1)R_L} > \text{maximum standoff ratio}$$

Often, the value of R_L must be selected to match the impedance of a particular load.

The effects of temperature must also be considered. Two important temperature effects are involved in the use of the emitter-follower stage.

First, the variation in β with temperature will change the loading and thus affect the frequency of oscillation. To minimize this temperature effect, $(\beta + 1)R_L$ should be much greater than the value of R_T. The values of C_T and R_T are chosen to provide the desired sawtooth frequency, as shown by the equations in Fig. 3-1. However, design should start with R_T, by making R_T a value between 0.1 and 0.2 times $(\beta + 1)R_L$. If this results in an impractical value for C_T, increase the value of R_T only as necessary to reduce the size of C_T to practical limits.

The second temperature effect results from the collector leakage current of the two-junction transistor, shown as I_{CO} in Fig. 3-2. Note that this current adds to the emitter leakage current I_{EO} of the UJT. Both leakage currents tend to increase the frequency with increasing temperature. The effect of leakage currents of frequency can be minimized by using a large value for C_T. If the NPN transistor is silicon, the effects of the two leakage currents can be neglected at temperatures below 100°C.

Some improvement in circuit operation can be achieved by using a PNP transistor as the emitter follower. With a PNP, the effective load resistance $(\beta + 1)R_L$, is in parallel with R_T so that the possibility of non-oscillation due to a low value of β or R_L is eliminated. Another advantage is that the I_{CO} of the transistor subtracts from the I_{EO} of the UJT so that some degree of temperature compensation is obtained. This is particularly true if a silicon PNP transistor is used.

3-1.2. Design Example of UJT Sawtooth Oscillator

Assume that the circuit of Fig. 3-1 is to provide a sawtooth output of approximately 5-V minimum into a 1-kΩ load. The available power source is 20 V, and a transistor with a β of 50 is to be used. The UJT has a maximum standoff ratio of 0.7.

The value of R_L is 1 kΩ to match the 1-kΩ load.

With R_L at 1 kΩ, and a β of 50, the value of $(\beta + 1)R_L$ is 51 kΩ.

With $(\beta + 1)R_L$ at 51 kΩ, the value of R_T should be between 5.1 kΩ and 10.2 kΩ. Assume a value of 5.1 kΩ.

Substituting these values for comparison against the maximum standoff ratio, we have

$$\frac{(50 + 1) \times 1000}{5100 + [(50 + 1) \times 1000]} = \frac{51{,}000}{56{,}100} \approx 0.9$$

Since 0.9 is greater than the maximum standoff ratio of 0.7, oscillation should be sustained with no difficulty.

With an available source voltage V_1 of 20 V, the V_0 output voltage should be approximately 6 V (20 × 0.3) which is greater than the required 5-V output. With 6 V across an R_L of 1 kΩ, there is 6 mA of current through the emitter follower. Of course, the emitter follower must be capable of dissipating this power plus whatever current is passing through the load.

With R_T at 5.1 kΩ, select the value of C_T to produce the desired frequency (or period) of operation (as shown by the equations in Fig. 3-1).

3-1.3. Improving Linearity of UJT Sawtooth Oscillators

For many applications, the sawtooth linearity obtained with the basic UJT relaxation oscillator is inadequate. Those UJTs with the *lowest standoff ratio* produce the most linear sawtooth waveforms. However, 10 per cent is about the best linearity that can be obtained with low standoff ratio UJTs.

A number of simple circuit techniques can be used to improve linearity of the sawtooth waveform from a UJT oscillator. Some practical circuits are shown in Fig. 3-3.

High voltage. Figure 3-3(a) shows the direct approach of using a higher supply voltage for charging the timing capacitor. This is an inexpensive method of improving linearity if the high voltage supply is available. The circuit of Fig. 3-3(a) has some disadvantage in that frequency is not as stable as with a single power supply.

Charging inductance. Figure 3-3(b) shows the use of a charging choke to maintain a constant charging current for charging the timing capacitor. This circuit requires a time constant for the charging circuit which is much greater than the period of oscillation, as shown by the equation. This condition usually produces an impractical size of L at oscillator frequencies of less than 1 kHz.

Collector characteristics. Figure 3-3(c) shows the use of the high output impedance of a common-base transistor to maintain a constant charging current for the capacitor. The values of R_T and C_T remain the same as for the basic circuit.

Two variations of a *bootstrap circuit* are shown in Figs. 3-3(d) and 3-3(e).

Zener bootstrap. In Fig. 3-3(d), a constant voltage is maintained across R_3 by Zener diode D_1 and the emitter-follower transistor amplifier stage so that the capacitor charging current is constant over the complete cycle. This circuit is quite economical because it makes double use of the transistor, both as a driver for the bootstrap circuit and as an output amplifier stage. Note that R_4 is returned to a negative voltage. If R_4 is grounded, the current flowing through D_1 would cause clipping at the bottom of the sawtooth waveform.

The values shown for C_1, R_1, R_2 and R_3 provide a sawtooth output at about 2 kHz. These values are selected on the same basis as for the basic circuit. The value of D_1 is 6 V, which is typical for a supply voltage in the 20- to 25-V range. The value of R_4 is chosen to match a given load impedance. However, a change in R_4 value will change the output voltage, all other factors remaining the same. Note that the frequency of the Fig. 3-3(d) circuit is somewhat dependent upon the supply voltage.

Capacitor bootstrap. In Fig. 3-3(e) the circuit uses a capacitor C_2 in place of the Zener diode. This variation permits the negative supply to be eliminated, thus making frequency largely independent of the supply voltage.

In each of the circuits shown in Figs. 3-3(a) to 3-3(e), the linearity is limited by the loading effect of the output stage so that it will not be possible to increase linearity beyond a certain value. This value is set by $(\beta + 1)R_L$.

RC integrator. The circuit of Fig. 3-3(f) shows a method of compensating for both the loading of the output stage and the variable charging current of the timing capacitor. Resistor R_3 and capacitor C_2 act as an integrating network of the waveform. By varying the values of R_3 and C_2, the output waveform can be made concave upward, concave downward, or linear. In a practical application, the circuit is assembled in breadboard form, using a potentiometer for R_3. The output is monitored on an oscilloscope, and the value of R_3 is adjusted to provide the most linear waveform.

Sec. 3-1　　　　　　　　　　　　　　　　　　　Sawtooth Oscillators　65

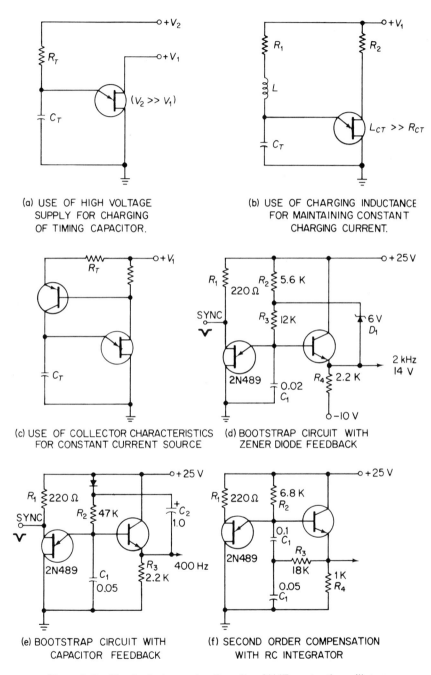

Figure 3-3　Circuits for improving linearity of UJT sawtooth oscillators (Courtesy General Electric)

3-2. MULTIVIBRATORS

All three types of transistors (two-junction, FET, and UJT) can be used to form multivibrator circuits. In this section we shall discuss two basic multivibrators, one which uses two-junction transistors (or FETs), and the other which uses a single UJT.

3-2.1. Basic Two-Transistor Multivibrator

Figure 3-4 is the working schematic of a basic multivibrator. The circuit uses a pair of two-junction transistors. FETs can also be used in a similar circuit. The circuit of Fig. 3-4 is of the *high-current type*. That is, the emitter resistor values are about half the collector resistor values, resulting in very high switching currents through the transistors. This requires transistors with a higher current capability (and higher power dissipation), but produces

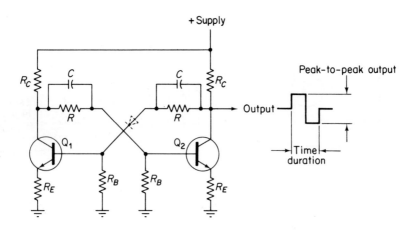

Output voltage (peak-to-peak) $\approx 0.6 \times$ supply $\approx$ operating point collector voltage

Q_1 and Q_2 power dissipated $\approx$ collector voltage $\times$ collector current

$R \approx 10 \times R_C$ Frequency $\approx \dfrac{1}{RC}$ Time duration $\approx RC$ time constant

Lowest collector voltage $\approx 0.3 \times$ supply

Maximum collector current $\approx \dfrac{\text{power dissipation limit of } Q_1 \text{ and } Q_2}{\text{lowest collector voltage}}$

$R_C \approx \dfrac{\text{supply} - \text{lowest collector voltage}}{\text{maximum collector current}}$

$R_E \approx 0.5 \times R_C$

$C \approx \dfrac{1}{\text{freq.} \times R}$

$R_B =$ see text

Figure 3-4   High-current multivibrator

a circuit with high-frequency stability. That is, the circuit will maintain frequency to about 1 part in 10^4, in spite of large changes in supply voltage.

The same circuit can be used as a *low-current type* where the collector resistors are about ten times the emitter-resistance value. This will require a lower current capability (and lower power dissipation) for the transistors, but will result in frequency changes with variations of the supply voltage. This is not always an undesirable condition. For example, in some telemetry circuits, a low-current type multivibrator is used as a voltage-to-frequency converter to transmit voltage information. However, for stability, the high-current type is far superior.

In either type, the output is a symmetrical square wave. That is, both the positive and negative portions of the cycle are of the same duration and amplitude. The output can be taken from either half of the circuit. Likewise, either half of the circuit can be triggered, if desired. The circuit can be either free-running, where the frequency is determined by the *RC* time constant, or the circuit can be driven by a trigger source.

3-2.2. Design Considerations for a Two-Transistor Multivibrator

The following specific design considerations should be considered. *Note that only the high-current type circuit is discussed.*

Output voltage. The peak-to-peak output voltage (or square wave) is about 0.6 times the supply voltage if the collector resistance is about twice the emitter resistance. For example, if the supply is 10 V, each collector will vary between about 3.5 V and 9.5 V, on alternate half-cycles.

Power dissipation. The power dissipated by each transistor can be approximated when the lowest collector voltage is multiplied by the collector current. For example, using the same 10-V supply, a 1-kΩ collector resistor, and a drop of 3 V, the current is $10 - 3$, or 7; 7/1000 or 7 mA. With 7 mA and 3 V, the transistor power dissipation is 21 mW.

Actually, this is minimum power dissipation (due to base current, etc.). To add a margin of safety, assume that the full supply voltage is applied with full current. In the above case, 7 mA $\times$ 10 V or 70 mW, is the maximum power dissipation required.

Worst case design. Since a multivibrator is a switching circuit, the principles of worst case design should be applied. In simple terms, this means that the base current used should be about three times the theoretical base current. Generally, base current is calculated on the basis of required collector current divided by gain. In switching circuits, use three times the calculated base current.

When the worst case design principle is applied, the transistor will always be over-driven. While this is undesirable in most linear circuits, it is required for switching circuits. Under normal operating conditions, a switching circuit is driven into full saturation, or full cutoff, in the *shortest possible time*.

Switching time. The total switching time of each transistor must be far less than the duration of one output cycle. Total switching time of a transistor is considered as the rise time, fall time, delay time, and storage time all added together.

As a rule, switching time must be less than 0.1 of the pulse duration (for a complete cycle). For example, if a transistor has a total switching time of 1 μS, the pulse duration should be at least 10 μS. Thus, the maximum multivibrator operating frequency is 100 kHz. Switching time information is found on the datasheets for switching transistors (or transistors that could be used in switching applications).

Symmetrical output. If both halves of the circuit are symmetrical, the output will be symmetrical. In some free-running circuits, a slight unbalance is introduced to ensure starting. A multivibrator is inherently stable, and may not start in the free-running configuration. An alternative method to start a free-running multivibrator is to connect a diode in series with either of the feedback resistances, as shown in dotted form in Fig. 3-4. The circuit will then be asymmetrical until it starts and reaches full operation.

Operating frequency. The operating frequency is determined by the time constant of the feedback RC, and is approximately equal to the reciprocal of the time constant. The exact frequency is difficult to calculate, since the transistor characteristics can affect the charge and discharge function.

Various combinations of R and C could be used to produce a given time constant (and thus a given frequency). However, the value of R should be approximately equal to ten times the value of the collector resistor. Capacitors of corresponding values should be selected to produce the desired frequency.

Bias relationships. The bias circuit can be calculated and tested on the basis of normal operating point, even though the circuit will never actually be there. A feedback signal will always be present, and the transistors will always be in a state of transition between full saturation and full cutoff.

With the capacitors removed, both halves of the circuit should be forward-biased, and the collector should be approximately 0.6 of the supply voltage. Likewise, the emitter should be about 0.2 (or less) of the supply voltage, and the base should be about +0.5 V (for NPN transistors) or negative (or PNP transistors) in relation to the emitter. Since the values of the collector, emitter, and feedback resistances are set by other circuit considerations, the value of R_B (base resistance) must be chosen to provide the necessary bias.

Sec. 3-2 Multivibrators 69

In practice, the values should be calculated, and the circuit assembled in experimental form, but with the capacitors omitted. The emitter, collector, and base should be checked for the desired voltage relationships. Then, if necessary, the value of R_B should be adjusted to produce the desired relationships. For a symmetrical output, the element voltages should be the same at both transistors.

3-2.3. Design Example of a Two-Transistor Multivibrator

Assume that the circuit of Fig. 3-4 is to provide a 12-V output at 7 kHz, operating as a free-running multivibrator. The available transistor has a total switching time of 1 μS, will dissipate 100 mW without a heat sink.

With a required output of 12 V (peak-to-peak), the supply must be 12/0.6, or 20 V. With a 20-V supply, each transistor will drop to about 7 V when fully saturated. This is a 13 V drop.

If the transistors are to be operated without heat sinks, the transistor power dissipation can not be exceeded. To allow some safety margin, assume that the dissipation limit is 90 mW instead of the rated 100 mW. With the collector at 7 V, and a maximum dissipation of 90 mW, the maximum allowable current is 0.090/7, or 13 mA.

With a 13-mA current and a 13-V drop (under maximum conditions) the collector resistors R_C should be 13/0.013, or 1 kΩ.

With 1 kΩ for the collector resistors, the emitter resistors R_E should be 500 Ω (510 Ω as the nearest standard), and the feedback resistors R_{feedback} should be 10 kΩ.

The value of the base resistors R_B should be calculated on the basis of operating point. Under the operating point conditions (capacitors removed from the circuit), the collectors should be at about 0.6 of the supply voltage, or at about 12 V, while the emitter should be slightly less than 0.2 of the supply voltage, or about 4 V. Under these conditions, there must be about 8 mA of collector current flowing, producing an 8-V drop across RC, and a 4-V drop across R_E.

With the emitters at about 4 V, the base should be at about 4.5 V ($+4.5$ V for the NPN shown). With 4.5 V at each base, and 12 V at each collector, there must be a 7.5-V drop across each feedback resistor. Since the feedback resistors are 10 kΩ, the current must be 0.75 mA to produce a 7.5-V drop.

The 0.75-mA current through the feedback resistors is a combination of base current, and the current through the base resistors R_B. The base current can be estimated on the basis of gain and required collector current. Assume that the rated gain is 120. Using half of this value for safety, and a required 8-mA collector current, the base current is 0.008/60 or 0.13 mA.

By subtracting the 0.13-mA base current from the 0.75-mA feedback resistor current, the base resistor R_B current is 0.75 $-$ 0.13, or 0.62 mA.

With base voltage of 4.5 V, and an R_B current of 0.62 mA, the R_B resistance values are 4.5/0.00062 or 7.3 kΩ (the nearest 10 per cent standard value is 6.8 kΩ).

In practice, the circuit should be assembled in breadboard form, omitting the capacitors, and using variable resistors for each R_B. Both R_B values should be adjusted as necessary for equal voltages at corresponding transistor elements. Then the capacitors can be connected, and the output waveform checked for correct frequency, waveform and amplitude. With the resistance ratios selected, the output amplitude should be approximately equal to the operating point collector voltage (or about 12 V in this case).

With a value of 10 kΩ for the feedback resistors, and a frequency of 7 kHz, the value of the feedback capacitors is 1/(10,000 × 7000), or 0.014 μF.

3-2.4. UJT Multivibrators

It is possible to use a single UJT as a multivibrator. The basic UJT multivibrator circuit is shown in Fig. 3-5. Note that the circuit is essentially the same as for the basic relaxation oscillator, except for the addition of R_2 and CR_1. Diode CR_1 is forward-biased when C is being charged. Charging time is set by R_1 and C in the normal manner. However, when C is discharging, CR_1 is reverse-biased (the anode of the diode is negative). As long as the capacitor is charged, the discharge current must flow through R_2, with the $R_2 C$ constant determining the discharge time.

As shown by the waveforms, the output at Base 2 is an approximate square wave, with the ON and OFF intervals separately controlled by R_1 and R_2. The time t_1 is the period for which the UJT is OFF, and t_2 is the period for which the UJT is ON, and CR_1 is reverse-biased.

A fairly accurate approximation of these times can be made by using the following equations

$$t_1 \approx R_1 C \ln \left(\frac{V_1 - V_E}{V_1 - V_P} \right)$$

$$t_2 \approx R_2 C \ln \left(\frac{V_1 + V_P - V_E}{V_1} \right)$$

where V_E is the emitter voltage measured at an emitter current $I_E = V_1(R_1 + R_2)/R_1 R_2$ and may be taken from the emitter characteristic curves.

However, a simplified rule is to *make R_2 approximately twice the value of R_1 when t_1 is to equal t_2*.

In a practical experimental circuit, calculate trial values for a given t_1 on the basis of a simple $R_1 C$ time constant. Then use a variable resistance for R_2, starting with a value twice that of R_1.

The basic UJT multivibrator may be coupled to a conventional transistor by means of a circuit such as shown in Fig. 3-6. In this circuit, the emitter-to-

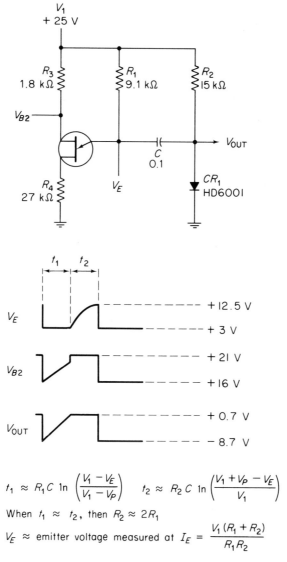

Figure 3-5 UJT multivibrator (Courtesy General Electric)

base diode of the transistor takes the place of the diode in the circuit of Fig. 3-5. The Fig. 3-6 circuit has the advantage that the load section is completely isolated from the timing section of the circuit. However, all of the timing values remain the same.

Note that in the circuit of Fig. 3-6, both Base 1 and Base 2 resistors are omitted. The Base 1 resistor is actually not required in either circuit, although

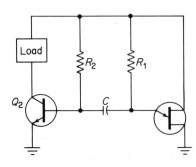

Figure 3-6 UJT multivibrator coupled to load through an NPN transistor

the addition of a Base 1 resistor to either circuit will minimize excessive emitter current when the UJT fires (if that is a problem with any particular application). The Base 2 resistor provides temperature compensation and can be used or omitted to fit a particular application. In the case of the Fig. 3-5 circuit, the Base 2 resistor R_3 is required since the output is developed across this resistor. In the Fig. 3-6 circuit, output is taken from transistor Q_2, and no temperature compensation is provided.

3-3. SCHMITT TRIGGERS

A Schmitt trigger is a bistable network which switches from one stable state to the other when an input signal varies above or below predetermined voltage levels. The Schmitt trigger is widely used as a voltage-level detector in analog and analog-digital systems. The Schmitt trigger is also used as a pulse-shaping circuit.

A JFET can be used in conjunction with a two-junction transistor to form a Schmitt trigger. Such an arrangement takes advantage of the JFET's high input impedance and the two-junction transistor's high-current capacity.

3-3.1. Design Considerations for Schmitt Triggers

Figure 3-7 shows a Schmitt trigger circuit configuration, using both the FET and two-junction transistors. Note that the input is applied at the FET gate, and the output is taken from the two-junction transistor collector. Assume, for the present analysis, that the FET Q_1 is removed from the circuit. The voltage level at point A, V_A, is sufficiently positive to turn on transistor Q_2; and the voltage level at point B, V_B, is clamped by the emitter-base junction of Q_2 to one diode drop below V_A (typically 0.5 to 0.7 V for a silicon transistor). With the FET Q_1 back in the circuit, the gate-source voltage V_{GS} of Q_1 is given by

$$V_{GS} = V_{IN} - V_B \qquad (3\text{-}1)$$

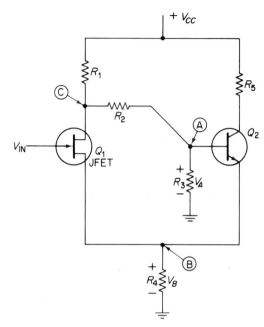

Figure 3-7 Schmitt trigger circuit using FET at input (Courtesy Texas Instruments)

The FET is nonconducting if V_{GS} is at least as negative as the pinch-off voltage V_P of the device. Hence, from Eq. (3-1), the FET is pinched off if

$$V_{IN} - V_B \leq V_P \qquad (3\text{-}2)$$

Circuit design is such that the inequality of Eq. (3-2) is valid for a zero-level input signal; output voltage of the circuit is low as Q_2 is turned on and Q_1 is turned off. As V_{IN} is made increasingly positive, a voltage level is reached where $V_{IN} - V_B$ becomes slightly more positive than V_P and the FET begins to conduct current.

The resulting voltage drop across R_1 causes Q_2 to partially turn off. This lowers the V_B level and Q_1 turns on even harder. The regenerative action causes the Schmitt trigger circuit to switch to the state where Q_1 is fully turned on, and Q_2 is at cutoff. Output voltage V_{OUT} is then at a relatively positive level.

As V_{IN} is reduced from a large positive voltage level toward ground potential, Q_1 begins to conduct less current. Eventually, a level of V_{IN} is reached at which the inequality of Eq. (3-2) is satisified. The circuit then switches to the state where Q_1 is turned off and Q_2 is fully on.

The quiescent states of the Schmitt trigger circuit correspond to the operating conditions where Q_1 is on and Q_2 is off, or where Q_1 is off and Q_2 is on. Therefore, the circuit has two stable stages, determined by the level of V_{IN}.

Open-loop gain. Regeneration occurs in the Schmitt trigger circuit when the open-loop gain of the circuit is equal to, or greater than, unity. This gain can be determined by opening the circuit at a convenient point and treating the resulting network as a linear amplifier. The circuit can be opened at the drain of Q_1, as shown in Fig. 3-8.

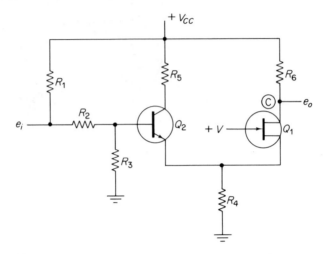

Figure 3-8 Network for obtaining open-loop gain of Schmitt trigger circuit (Courtesy Texas Instruments)

An input signal e_i is applied to resistor R_2. The output signal e_o appears at point C. Transistor Q_2 is biased in the active region by R_1 connected to the supply voltage. A d-c voltage $+V$ is applied to the gate of Q_1 in order to bias Q_1 in the active region. An expression for open-loop gain can be described by

$$A_V \approx \frac{R_1 R_3}{\left[\dfrac{R_2 R_3}{R_E(h_{FE}+1)} + R_2 + R_3\right]\left(\dfrac{1 + R_1 Y_{os}}{Y_{fs} + Y_{os}}\right)} \qquad (3\text{-}3)$$

where:

A_V is open-loop voltage gain

Y_{os} is small-signal common-source output admittance of Q_1

Y_{fs} is small-signal common-source forward transfer admittance of Q_1

h_{FE} is d-c current gain of Q_2

R_E is equal to $(1 + Y_{os}R_1)R_4/[R_4(Y_{fs} + Y_{os}) + R_1 Y_{os} + 1]$

Parameters Y_{os} and Y_{fs} in Eq. (3-3) are dependent upon d-c operating

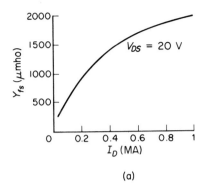

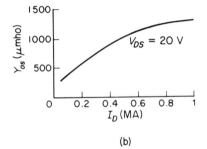

Figure 3-9 Measured plots of y_{fs} and y_{os} for a typical FET

(a)

(b)

levels of the FET. Figure 3-9 shows measured plots of these two parameter for an FET. At extremely low current levels, the small value of Y_{fs} prevents regeneration in the Schmitt trigger circuit. However, as current increases in the FET, Y_{fs} becomes large enough for the open-loop gain to exceed unity.

Equation (3-3) can be reduced to fewer terms by using the relationships $R_1 Y_{os} \ll 1$ and $Y_{fs} \gg Y_{os}$. This allows Eq. (3-3) to be simplified to

$$A_V \approx \frac{R_1 R_3 Y_{fs}}{\dfrac{R_2 R_3}{R_E(h_{FE}+1)} + R_2 + R_3} \tag{3-4}$$

The above expression is used to determine Y_{fs} for unity circuit gain. From a $Y_{fs} - Y_{GS}$ plot of the type shown in Fig. 3-10, the corresponding V_{GS} level can be determined.

The validity of Eq. (3-4) was checked by breadboarding the circuit of Fig. 3-11. The FET (the characteristics of which are shown in Figs. 3-9 and 3-10) is used in the circuit of Fig. 3-11. For a given level of I_D, a corresponding value for Y_{fs} is obtained from 3-11(a). This Y_{fs} level is then substituted into Eq. (3-4) together with resistor and h_{FE} values. Figure 3-11(b) shows

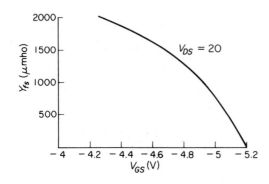

Figure 3-10 Measured plot of y_{fs} versus V_{GS} for a typical FET

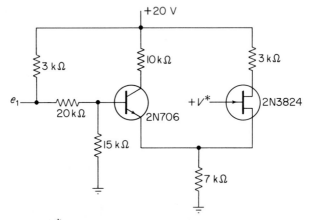

*+V is adjusted to give desired level of I_D

(a)

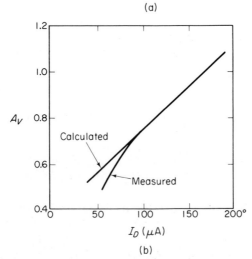

(b)

Figure 3-11 Test circuit for determining open-loop gain, and plots of measured and calculated open-loop gain

measured and calculated plots of A_V versus I_D. These plots show that Eq. (3-4) gives a close approximation to open-loop voltage gain.

Trigger-voltage levels. The input voltage level at which the Schmitt trigger circuit changes from the low-output voltage state to the high-output state is referred to as V_{ON}. The input voltage level at which the circuit switches back to the low-output voltage state is designated V_{OFF}.

Consider that input voltage V_{IN} (to the circuit of Fig. 3-7) is at a low potential, and that Q_2 is conducting. Transistor Q_2 may be either in saturation or in the active region, depending upon the circuit design. If Q_2 is saturated, the circuit does not behave as a linear amplifier, and a small voltage change at point C has no effect upon collector current of Q_2. In order for collector current of Q_2 to change, drain current of Q_1 must increase to the level where the voltage drop across R_1 causes Q_2 to turn off slightly. However, if Q_2 is not saturated, the circuit is a linear amplifier at any level of drain current, and regeneration occurs at a lower level of V_{IN}. Thus, Q_2 is maintained out of saturation for the following analysis:

For Q_1 is cutoff in Fig. 3-7, V_B is given by

$$V_B = V_A - V_{BE(ON)} \tag{3-5}$$

where: $V_{BE(ON)}$ is the base-emitter forward voltage drop of Q_2. In the last expression, V_A can be described by

$$V_A = \frac{[V_{CC} - I_B(R_1 + R_2)]R_3}{R_1 + R_2 + R_3} \tag{3-6}$$

Substitute the right side of Eq. (3-5) for V_B in Eq. (3-1). Also from Eq. (3-5), substitute V_{ON} for V_{IN} in Eq. (3-1). Solving the resulting expression for V_{ON} gives

$$V_{ON} = V_{GS} + V_A - V_{BE(ON)} \tag{3-7}$$

The latter equation gives the value of V_{ON} in terms of circuit parameters. Voltage V_{GS} is at the level which makes open-loop circuit gain equal unity.

When the Schmitt trigger circuit of Fig. 3-7 switches to the state where Q_1 is ON and Q_2 is OFF, base-emitter voltage of Q_2 is given by

$$V_{BE(OFF)} = \frac{(V_{CC} - I_D R_1)R_3}{R_1 + R_2 + R_3} - I_D R_4 \tag{3-8}$$

where $V_{BE(OFF)}$ is off-state voltage of Q_2, and I_D is the drain current of Q_1.

The above level of $V_{BE(OFF)}$ is now more negative than the base-emitter turn-on voltage of Q_2. As V_{IN} is decreased, I_D decreases and $V_{BE(OFF)}$ becomes more positive. In order for Q_2 to begin conducting, the level of $V_{BE(OFF)}$ must increase to approximately 0.5 V (assuming that Q_2 is a low-power silicon transistor).

3-3.2. Design Example of Schmitt Triggers

Assume that the circuit of Fig. 3-7 is to be used as a Schmitt trigger. The circuit conditions are: $V_{ON} = 2$ V, $V_{OFF} = 1.8$ V, $V_{CC} = +20$ V. The FET is a 2N3824, having a V_P of approximately 5.15 V and characteristics similar to those shown in Figs. 3-9 and 3-10. A 2N706 transistor is used for Q_2.

Collector current I_C and R_5. The product $I_C R_5$ determines voltage swing at the output. When Q_2 is conducting, the magnitude of V_{CC} is dropped across the series combination of R_5, Q_2, and R_4. At this point in the analysis, none of the above three voltage drops is known.

The voltage drop across R_4 can be decribed by combining Eqs. (3-5) and (3-7), and solving the resulting expression for V_B; this yields

$$V_B = V_{ON} - V_{GS} \tag{3-9}$$

When Q_1 turns ON, V_{GS} is slightly less negative than V_P (assume a V_{GS} of -5 V). Under these conditions, V_B is not larger than $2 - (-5)$, or 7 V.

At least 1 V should be maintained across $Q_2 (V_{CE})$ in order to assume that Q_2 does not saturate. Voltage swing across R_5 can be seen as large as $20 - 7 - 1$, or 12 V.

Select a value of 10 V for the product $I_C R_5$. This gives a V_{CE} level of 2 V for the ON state of Q_2. The choice of values for I_C and R_5 is dependent largely upon capacitive loading at the circuit output, and any limitation upon fall time of the output voltage waveform (a lower value of R_5 will lower fall time). However, since this example has no specification for fall time, choose R_5 to be 10 kΩ. This given an I_C level of 1 mA.

Resistors R_3 and R_4. For a collector current of 1 mA, V_{BE} of the 2N706 is measured to be 0.67 V. The value of V_{GS} to be used in the following equations is that value which will make open-loop gain equal to unity. However, since R_1, R_2, and R_3 are not known at this point, it is not possible to use Eq. (3-4) to determine the level of Y_{fs} (and, consequently, V_{GS}) to give a Y_{fs} of 500 μmhos (about -5 V). Substitution of the V_{GS} value (-5 V), the V_{BE} level (0.67 V), and the V_{ON} level (2 V) into the following equation yields a V_A of 7.67.

$$\begin{aligned} V_A &= V_{ON} - V_{GS} + V_{BE} \\ &= 2 - (-5) + 0.67 = 7.67 \end{aligned} \tag{3-10}$$

With V_A at 7.67, the voltage at point B (or V_B) is 7 V, since V_B is $V_A - V_{BE}$.

The value of R_4 can be determined by

$$R_4 = \frac{V_B}{I_C}$$

$$= \frac{7}{0.001} \quad (3\text{-}11)$$

$$= 7 \text{ k}\Omega$$

Figure 3-12 shows the input circuit to the base of Q_2. For Q_2 turned off, and neglecting the small collector-base reverse leakage current I_B, voltage V_A is given by

$$V_A = I_O R_3 \quad (3\text{-}12)$$

In order to maintain V_A at a level which is relatively independent of small changes in base current, let I_O be 10 times the value of I_B. Common-emitter d-c current gain h_{FE} of the particular 2N706 transistor used here has a value of 82 at 1 mA of collector current, and a V_{BE} of 0.67; 1 ma/82 gives a I_B level of 0.0122 mA. By making I_O 10 times I_B, the value of I_O is 0.122 mA. With V_A at 7.67 and I_O at 0.122, Eq. (8-12) shows that R_3 is $7.67/0.122 = 62.8$ kΩ.

Resistors R_1 and R_2. Current I_O in the circuit of Fig. 3-12 can be described by

$$I_O = \frac{V_{CC}}{1.1(R_1 + R_2) + R_3} \quad (3\text{-}13)$$

All terms except R_1 and R_2 are known at this point. Equation (3-8) also contains the terms R_1 and R_2. Equations (3-8) and (3-13) can be solved simultaneously for R_1 to give

$$R_1 = \frac{V_{CC}}{I_D} - \frac{(V_{BE(OFF)} + I_D R_4)(V_{CC} + 0.1 I_D R_3)}{1.1 I_D I_O R_3} \quad (3\text{-}14)$$

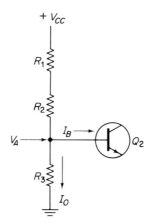

Figure 3-12 Input to base of Q_2
(Courtesy Texas Instruments)

All terms of Eq. (3-14) are known, except $V_{BE(OFF)}$ and I_D. $V_{BE(OFF)}$ is assumed to be 0.5 V (which is logical for a silicon transistor). The value of I_D should be that current which is produced (by V_{IN}) at the V_{OFF} level of 1.8 V.

A graphical procedure is used to determine I_D for a given level of V_{IN}. A portion of the FET transfer curve is shown in Fig. 3-13, together with a load line having a slope of $1/R_4$. This load line intersects the horizontal axis at the V_{OFF} level of 1.8 V. The load line also intersects the I_D curve at about 0.875 mA. Thus, an I_D of 0.875 is flowing when the gate of Q_1 is at 1.8 V (the V_{OFF} level). Using Eq. (3-14), R_1 is found to be 4.2 kΩ.

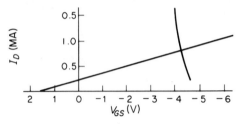

Figure 3-13 Transfer curve and load line for design example (Courtesy Texas Instruments)

With V_{CC}, I_O, R_1, and R_3 known, Eq. (3-13) can be rearranged to find R_2 as follows

$$R_2 = \frac{\frac{V_{CC}}{I_O} - R_3}{1.1} - R_1 \qquad (3\text{-}15)$$

$$= \frac{\frac{20}{0.122 \text{ mA}} - 62.8 \text{ k}\Omega}{1.1} - 4.2 \text{ k}\Omega$$

$$= 87.8 \text{ k}\Omega$$

4 PHOTOTRANSISTOR CIRCUITS

A phototransistor is a device used for controlling current flow with light. Basically, any transistor will function as a phototransistor if the chip is exposed to light. However, certain design techniques are required to make the best use of the phototransistor effect.

The circuits using phototransistors also require special design techniques. The circuit designer must supplement his conventional circuit knowledge with the terminology and relationships of *optics* and *radiant energy*. This chapter provides the information necessary to supplement that knowledge.

The chapter starts with a brief review of phototransistor theory and characteristics, followed by a discussion of irradiance, illuminance, and optics (concentrating on their significance to phototransistor circuit design). Both low-frequency/steady-state design and high-frequency design problems are considered. Use of the basic design information is then demonstrated with typical phototransistor circuit design examples.

4-1. PHOTOTRANSISTOR THEORY

Phototransistor operation is a result of the photo-effect in semiconductors. Light of a proper wavelength will generate hole-electron pairs (carriers) within the transistor, and an applied voltage will cause these carriers to move, thus causing a current to flow. The intensity of the applied light will determine the number of carrier pairs generated, and thus the magnitude of the resultant current flow.

In a phototransistor, the generation of carriers takes place in the vicinity of the collector-base junction. As shown in Fig. 4-1, for an NPN device, the

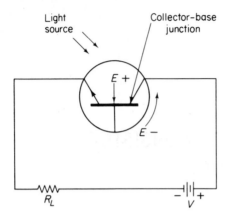

Figure 4-1 Photo-generated carrier movement in a photo-carrier

photo-generated holes will gather in the base. In particular, a hole generated in the base will remain there, while a hole generated in the collector will be drawn into the base by the strong field at the junction. The same process will result in electrons tending to accumulate in the collector. However, a charge will not accumulate at the junction. Instead, the charge will try to distribute evenly throughout the bulk regions. Consequently, holes will diffuse across the base region in the direction of the emitter junction.

When the holes reach the junction, they will be injected into the emitter. This, in turn, will cause the emitter to inject electrons into the base. Since the emitter injection efficiency is much larger than the base injection efficiency, each injected hole will result in many injected electrons.

At this point, normal transistor action occurs. The emitter-injected electrons travel across the base and are drawn into the collector. There, the electrons combine with the photo-induced electrons in the collector to appear as the terminal collector current.

Since the actual photogeneration of carriers occurs in the collector-base region, the larger the area of this region, the more carriers generated. Thus, as shown in Fig. 4-2, the phototransistor is designed to expose a large area to the light.

4-2. PHOTOTRANSISTOR STATIC CHARACTERISTICS

A phototransistor can be either a two-lead or a three-lead device. In the three-lead form, the base is made electrically available, and the device may be used as a standard two-junction transistor, with or without the additional capability of sensitivity to light. In the two-lead form, the base is not electrically available, and the transistor can only be used with light as an input. In most applications, the only drive to the transistor is light, and so the two-lead version is more prominent.

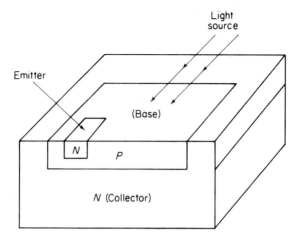

Figure 4-2 Typical double-diffused phototransistor structure

The phototransistor can be modeled as a two-lead device as shown in Fig. 4-3. In this circuit, current generator (I_{GEN}) represents the photo-generated current. The remaining elements of Fig. 4-3 should be recognized as the component distribution in the hybrid-pi transistor model. Note that the mode of Fig. 4-3 indicates that under dark conditions, I_{GEN} is zero and so

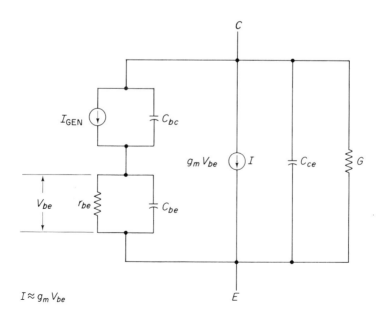

$I \approx g_m V_{be}$

Figure 4-3 Floating base approximate model of phototransistor

V_{be} is zero. This means that the terminal current I (approximately equal to $g_m V_{be}$) is also zero.

In reality, there is a thermally generated leakage current I_O which shunts I_{GEN}. This current I_{CEO} is typically on the order of 10 nA at room temperature and may be neglected (in most cases).

As a three-lead device, the model in Fig. 4-3 need only have a resistance r_b connected to the junction of C_{bc} and C_{be}. The other end of this resistance is the base terminal. Since the three-lead phototransistor is less common than the two-lead version, the only advantages of having the base leads available are to stabilize the operation of the device for wide temperature variations, or to use the base for unique circuit purposes.

Mention is often made of the ability to optimize a phototransistor's sensitivity by using the base. The idea is that the device can be electrically biased to a collector current at which h_{fe} (or β) is maximum. However, the introduction of any impedance into the base results in a net decrease in photosensitivity. This is similar to the effect noticed when I_{CEO} is measured for a transistor and found to be greater than I_{CER}. The base-emitter resistor shunts some current around the base-emitter junction, and the shunted current is never multiplied by h_{fe}. When the phototransistor is biased to peak h_{fe}, the magnitude of base impedance is low enough to shunt an appreciable amount of photo current around the base-emitter. The result is actually a lower device sensitivity than found in the open base mode.

4-2.1. Spectral Response of Phototransistors

By definition, a phototransistor is sensitive to light of particular wavelengths. In practice, however, response is found for a range of wavelengths. Figure 4-4 shows the normalized response for a typical phototransistor series (Motorola MRD devices) and indicates that peak response occurs at a wavelength of 0.8 μm (micrometers). The warping in the

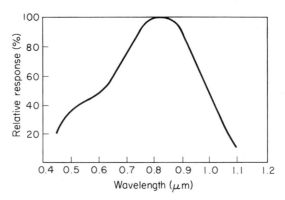

Figure 4-4 Constant energy spectral response for Motorola MRD series phototransistors (Courtesy Motorola)

response curve in the vicinity of 0.6 μm results from interference in the oxide layer covering the transistor surface.

4-2.2. Radiation Sensitivity of Phototransistors

The absolute response of an MRD450 phototransistor to radiation is shown in Fig. 4-5. This response is standardized to a tungsten source operating at a color temperature of 2870°K (color temperature is expressed in degrees Kelvin). As the following discussions will show, the transistor sensitivity is quite dependent on the *color temperature* of the light source.

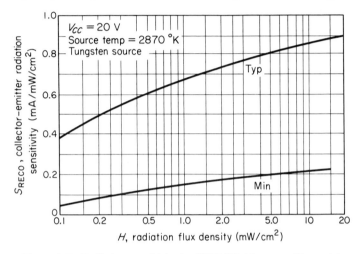

Figure 4-5 Radiation sensitivity for MRD450 (Courtesy Motorola)

4-3. RADIATION AND ILLUMINATION SOURCES

The effect of a radiation source on a phototransistor is dependent on the transistor spectral response, and the spectral distribution of energy from the source. When discussing such energy, two related sets of terminology are available. The first is *radiometric* which is a physical system; the second is *photometric* which is a physiological system.

The photometric system defines energy relative to the visual effect. As an example, light from a standard 60 W bulb is certainly visible and, as such, has finite photometric quantity. Radiant energy from a 60 W resistor is not visible and has zero photometric quantity. However, both the bulb and resistor have finite radiometric quantity.

The defining factor for the photometric system is the *spectral response curve of a standard observer*. This is shown in Fig. 4-6 and is compared with the spectral response of the MRD series. The defining spectral response of the radiometric system can be imagined as unit response for all wavelengths. A comparison of radiometric and photometric terminology is given in Fig. 4-7.

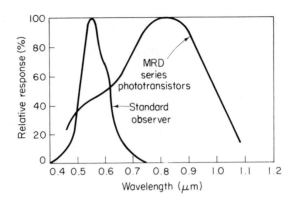

Figure 4-6 Spectral response for standard observer and MRD series phototransistors (Courtesy Motorola)

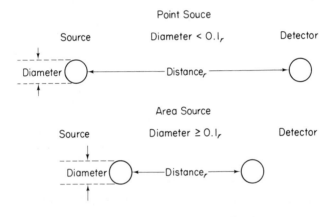

Radiometric and Photometric Terminology

Description	Radiometric	Photometric
Total flux	Radian flux, P, in watts	Luminous flux, F, in lumens
Emitted flux density at a source surface	Radiant emittance, W, in watts/cm^2	Luminous emittance, L, in lumens/ft^2 (foot-lamberts) or lumens/cm^2 (lamberts)
Source intensity (area source)	Radiance, B_r, in (watts/steradian)/cm^2	Luminance, B_L, in (lumens/steradian)/ft^2
Source intensity (point source)	Radiant intensity, I_r, in watts/steradian	Luminous intensity, I_L, in lumens/steradian (candela)
Flux density incident on a receiver surface	Irradiance, H, in watts/cm^2	Illuminance, E, in lumens/ft^2 (footcandle)

Figure 4-7 Comparison of radiometric versus photometric terminology, and definitions of point and area sources

There exists a relationship between the radiometric and photometric quantities such that at a wavelength of 0.55 μm (the wavelength of peak response for a standard observer), 1 W of radiant flux is equal to 680 lm (lumens) of luminous flux.

The photometric effect of a radiant source can often be measured directly with a photometer. Unfortunately, most phototransistors are specified for use with the radiometric system. Therefore, it is often necessary to convert photometric source data, such as the candle power rating of an incandescent lamp to radiometric data.

4-3.1. Geometric Considerations of Light Sources

In the design of electro-optic systems, the geometrical relationships are of prime concern. A source will effectively appear as either a *point source*, or an *area source*, depending upon the relationship between the size of the source and the distance between the source and the detector.

Point sources. A point source is defined as one for which the source diameter is less than ten per cent of the distance between the source and the detector. Figure 4-8 shows a point source radiating uniformly in every direction. Figure 4-8 also lists the design relationships for a point source in terms

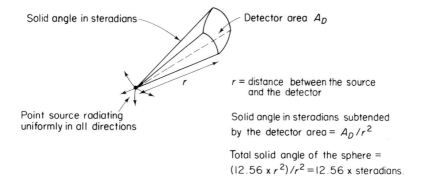

Point Source Relationships

Description	Radiometric	Photometric
Point source intensity	I_R, watts/steradian	I_L, lumens/steradian
Incident flux density	Irradiance $H = I_r/r^2$ watts/distance^2	Illuminance $E = I_L/r^2$ lumens/distance^2
Total flux output of point source	$P = 12.56\, I_r$ watts	$F = 12.56\, I_L$ lumens

Figure 4-8 Point source relationships

of both radiometric and photometric quantities. Figure 4-8 is based on the assumption that the photodetector is aligned such that its surface area is tangent to the sphere with the point source at its center. It is possible that the plane of the detector can be inclined at an angle from the tangent plane. Under this condition, as shown in Fig. 4-9, the irradiance (H) and illuminance (E) change (in proportion to the angle).

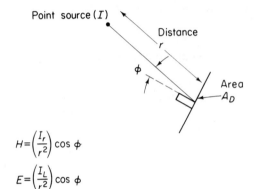

$$H = \left(\frac{I_r}{r^2}\right) \cos \phi$$

$$E = \left(\frac{I_L}{r^2}\right) \cos \phi$$

Figure 4-9 Irradiance H and illuminance E when detector is not normal to source direction

Area sources. When the source has a diameter greater than 10 per cent of the separation distance, it is considered to be an area source. This situation is shown in Fig. 4-10. Figure 4-11 lists the design relationships for an area source.

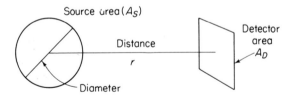

Diameter greater than 10% of distance

Figure 4-10 Area source geometry

A special case that deserves some consideration occurs when the diameter of the source is much greater than the separation distance (that is, when the detector is quite close to the source). Under these conditions, the emitted and incident flux densities are equal, and the total incident energy is approximately the same as the total radiated energy. When this occurs, *unity coupling* is said to exist between source and detector.

Special case where diameter/2 is greater than distance
(detector is quite close to source)

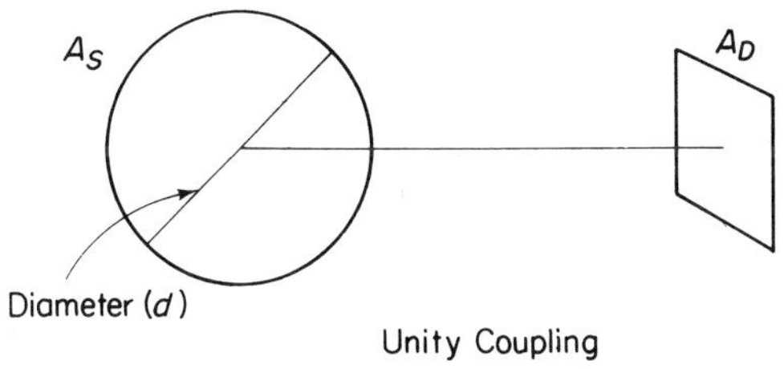

$$H = \frac{B_r A_S}{r^2 + \left(\frac{d}{2}\right)^2} \approx \frac{B_r A_S}{\left(\frac{d}{2}\right)^2}$$

$$A_S = 3.14 \left(\frac{d}{2}\right)^2$$

$$H \approx B_r \, 3.14 = W$$

Unity Coupling

Normal Design relationships for an Area Source

Description	Radiometric	Photometric
Source intensity	B_r watts/cm^2/steradian	B_L, lumins/cm^2/steradian
Emitted flux density	$W = 3.14 \, B_r$, watts/cm^2	$L = 3.14 B_L$, lumens/cm^2
Incident flux density	$H = \dfrac{B_r A_S}{r^2 + \left(\frac{d}{2}\right)^2}$, watts/cm^2	$E = \dfrac{B_L A_S}{r^2 + \left(\frac{d}{2}\right)^2}$, lumens/cm^2

Figure 4-11 Normal and special-case design relationships for an area source

4-3.2. Lens Systems for Phototransistors

A lens can be used with a photodetector to effectively increase the irradiance on the detector. As shown in Fig. 4-12(a), the *irradiance on a target surface for a point source of intensity I is*: irradiance = intensity/separation distance^2. In Fig. 4-12(b), a lens has been placed between the source and detector. If the lens radius is greater than the detector radius, the lens provides an increase in incident irradiance on the detector. The approximate gain of the lens system can be found by: lens system gain = $0.9 \times$ (lens radius/detector radius)^2.

It should be pointed out that arbitrary placement of a lens may be more harmful than helpful. That is, a lens system must be carefully planned to be effective.

For example, the MRD300 phototransistor contains a lens which is effective when the input is in the form of parallel rays (as approximated by a uniformly radiating point source). If a lens is introduced in front of the MRD300 as shown in Fig. 4-13, the lens will provide a non-parallel ray input to the transistor lens. Thus, the net optical circuit will be misaligned. The net irradiance on the phototransistor chip may in fact be less than without the external lens.

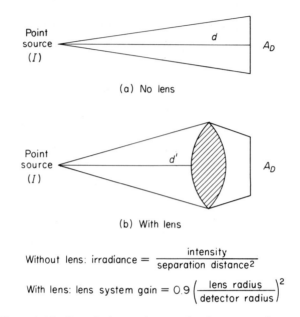

$$\text{Without lens: irradiance} = \frac{\text{intensity}}{\text{separation distance}^2}$$

$$\text{With lens: lens system gain} = 0.9 \left(\frac{\text{lens radius}}{\text{detector radius}}\right)^2$$

Figure 4-12 Use of a lens to increase irradiance on a detector

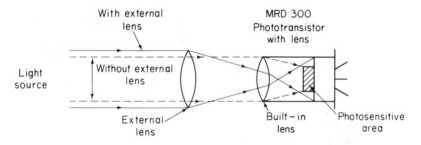

Figure 4-13 Possible misalignment due to arbitrary use of external lens

The arrangement of Fig. 4-14 shows an effective lens system. Lens 1 converges the energy incident on its surface to lens 2 which reconverts this energy into parallel rays. The energy entering the phototransistor lens as parallel rays is the same (neglecting losses) as that entering lens 1. Another way of looking at this is to imagine that the phototransistor surface has been increased to a value equal to the surface area of lens 1.

4-3.3. Fiber Optics for Phototransistors

Another technique for getting the best coupling between source and detector is to use a fiber bundle to link the phototransistor with the light source. Operation of fiber optics is based on the principle of *total internal*

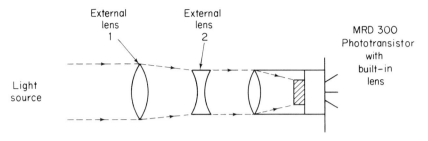

Phototransistor surface ≈ surface of external lens 1

Figure 4-14 Effective use of external optics with MRD300 phototransistor

reflection. Figure 4-15 shows the application of this principle to fiber optics. A glass fiber of refractive index n is clad with a layer of glass of lower refractive index n'. A ray of light entering the end of the fiber will be refracted as shown. (Light is refracted when it passes through the interface of two dissimilar materials.)

If, after refraction, the light approaches the glass interface at an angle greater than the "critical angle" in Fig. 4-15, the light will be reflected within the fiber. As with any optic system, the angle of reflection equals the angle of incidence. Thus, the light ray will bounce down the fiber and emerge, refracted, at the exit end. For total internal reflection to occur, a light ray must enter the fiber within the half-angle, as shown (outside of the "critical angle").

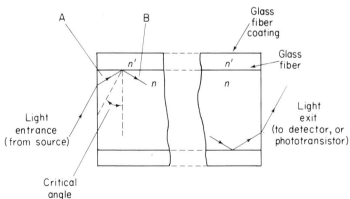

Figure 4-15 Refraction in an optical fiber used between a light source and a phototransistor

Once the light ray is within the fiber, the ray will suffer some attenuation. For typical glass fibers, an *absorption rate from 5 to 10 per cent per foot* is a reasonable guideline. Also, there is an entrance and exit loss at the ends of the fiber which typically result in about 30 per cent loss. As an example, an illuminance at the source end of a 3-ft fiber bundle will appear at a level of about 50 per cent at the detector.

4-3.4. Tungsten Lamps for Phototransistors

Tungsten lamps are often used as radiation sources for photodetectors. The radiant energy of these lamps is distributed over a broad band of wavelengths. Since the eye and the phototransistor exhibit different wavelength-dependent response characteristics, the effect of a tungsten lamp will be different for both. The spectral output of a tungsten lamp is very much a function of *color temperature*.

Color temperature of a lamp is the temperature required by an ideal blackbody radiator to produce the same visual effect as the lamp. At low color temperatures, a tungsten lamp emits very little visible radiation. However, as color temperature is increased, the response shifts toward the visible spectrum.

Figure 4-16 shows the spectral distribution of tungsten lamps as a function of color temperature. The lamps are operated at a constant wattage and the response is normalized to the response of 2800°K. For comparison, the spectral response of both the standard observer and the Motorola MRD phototransistors are also plotted.

Effective irradiance. Although the sensitivity of a photodetector to an

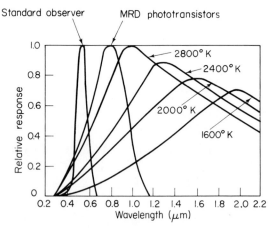

Figure 4-16 Radiant spectral distribution of tungsten lamp as a function of color temperature

illuminant source is frequently provided, the sensitivity to an irradiant source is more common. Thus, it is advisable to carry out design work in terms of irradiance. However, since the spectral response of a source and detector are, in general, not the same, a response integration must be performed.

Graphical integration has been performed for the MRD phototransistors for several values of lamp color temperature. The results are given in Fig. 4-17 in terms of ratios.

Figure 4-17 provides the irradiance ratio H_E/H versus color temperature. As shown, a tungsten lamp operating at 2600°K is about 23.6 per cent effective on the MRD series phototransistors. That is, if the broadband irradiance of such a lamp is measured at the detector and found to be 20 mW/cm², the transistor will effectively see $0.236 \times 20 = 4.72$ mW/cm².

The specifications for the MRD phototransistor series include the correction for effective irradiance. For example, the MRD450 is rated for a typical sensitivity of 0.8 mA/mW/cm². This specification is made with a tungsten source operating at 2870°K and provides for an irradiance at the transistor of 5 mW/cm². Note that this will result in a current flow of 4 mA (5 mA $\times$ 0.8). However, from Fig. 4-17, the effective irradiance is: $0.255 \times 5 = 1.28$ mW/cm².

By using this value of effective irradiance, and the typical sensitivity rating current flow of 4 mA, the *monochromatic irradiance* is: 4 mA/1.28 mW/cm² = 3.13 mA/mW/cm².

Now, as discussed previously, an irradiance of 20 mW/cm² produces an effective irradiance of 4.72 mW/cm². (That is, an irradiance of 20 mW/cm² looks like a monochromatic irradiance of 4.72 mW/cm².) Thus, the resultant current flow is: $3.13 \times 4.72 = 14.0$ mA.

An alternate approach is provided by Fig. 4-18 which shows the *relative*

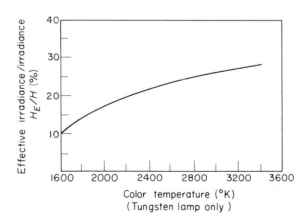

Figure 4-17 Irradiance ratio versus color temperature for MRD phototransistors (Courtesy Motorola)

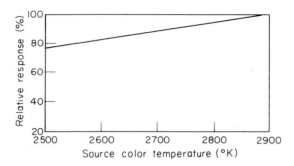

Figure 4-18 Relative response of MRD phototransistors versus color temperature

response as a function of color temperature. As shown, the response is down to 83 per cent at a color temperature of 2600°K. The specified typical response for the MRD450 at 20 mW/cm² for a 2870°K tungsten source is 0.9 mA/mW/cm². The current flow at 2600°K and 20 mW/cm² is: 0.83 × 0.9 × 20 = 14.9 mA.

Determination of color temperature. It is likely than an electronic circuit designer will not have the capability to measure color temperature. However, with a voltage measuring capability, a reasonable approximation of color temperature may be obtained.

Figure 4-19 shows the classical variation of lamp current, candle power, and lifetime for a tungsten lamp as a function of applied voltage. Figure 4-20 shows the variation of color temperature as a function of the ratio: MSCP/WATT; where MSCP is the mean spherical candle power at the lamp

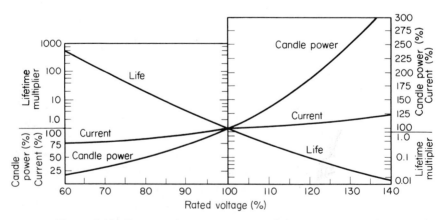

Figure 4-19 Tungsten lamp parameter variations versus variations about rated voltage (Courtesy Motorola)

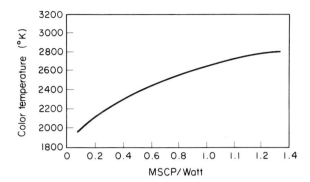

Figure 4-20 Color temperature versus candle power/power ratio (MSCP/Watt ratio) (Courtesy Motorola)

operating point, and WATT is the lamp current-voltage produced at the operating point.

As an example, assume that a type 47 indicator lamp is used as a source for a phototransistor. To extend the lifetime, the lamp is operated at 80 per cent of rated voltage. A type 47 lamp is rated at 6.3 V, 150 mA and an MSCP of 0.52.

From Fig. 4-19, using 80 per cent of rated voltage, the rated current is about 86 per cent. Thus, the operating point current is about 150 mA × 0.86 = 129 mA.

Again from Fig. 4-19, using 80 per cent of rated voltage, the rated candle power is about 50 per cent. Thus, the operating point candle power is about 0.52 × 50 per cent = 0.26 candle power (MSCP = 0.26).

The rated voltage is 6.3 V. 80 per cent of rated voltage is 5.04 V. WATTs is found when the operating point current of 129 mA is multiplied by the operating point voltage of 5.05 V, or 0.65 W.

To find the ratio of MSCP/WATT, substitute the values 0.26/0.65 for a ratio of 0.4. From Fig. 4-20, with a ratio of 0.4, the color temperature is approximately 2300°K.

Geometric considerations. The candle power ratings on most lamps are obtained by measuring the total lamp output in an integrating sphere and dividing by the unit solid angle. Thus, the rating is an average, or MSCP. However, a tungsten lamp cannot radiate uniformly in all directions, and the candle power varies with the lamp orientation.

Figure 4-21 shows the radiation pattern for a typical frosted tungsten lamp. The maximum radiation occurs in the horizontal direction for a base-down or base-up lamp. The broken curve simulates the output of a uniform radiator, and contains the same area as the lamp polar plot. The curve indicates that the lamp horizontal output is about 1.33 times the rated MSCP, whereas the vertical output opposite the base is 0.48 times the rated MSCP.

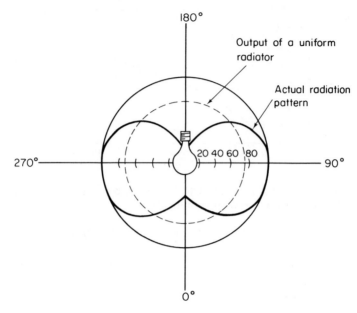

Figure 4-21 Typical radiation pattern for frosted tungsten lamp

The actual polar variation for a lamp will depend on a variety of physical features such as filament shape, size and orientation as well as the solid angle intercepted by the base with respect to the center of the filament.

If the lamp output is given in *horizontal candle power* (HCP), a fairly accurate calculation can be made with regard to illuminance on a receiver.

Another form of rating is *beam candle power*, which is provided for lamps with reflectors. In any lamp radiation measurement system, the rating may be given in lumens/steradians (lm/sr) or candle power.

4-3.5. Solid-State Light Sources for Phototransistors

In contrast with the broadband source of radiation provided by the tungsten lamp, solid-state sources provide relatively narrow band energy. The gallium arsenide (GaAs) light-emitting-diode (LED) has spectral characteristics which make it a favorable mate for use with silicon phototransistors. LEDs are available for several wavelengths, as shown in Fig. 4-22.

The GaAs diode and the MRD phototransistor series are particularly compatible. An efficiency ratio of effective irradiance/absolute irradiance (H_E/H) of about 0.9, or 90 per cent is possible when the GaAs diode and MRD phototransistor are used together. That is, an irradiance of 4 mA/cm² from an LED (of the GaAs type) will appear to the phototransistor as 3.6 mW/cm². This means that a typical GaAs LED is about 3.5 times as effective as a tungsten lamp at 2870°K. Thus, the typical sensitivity for the

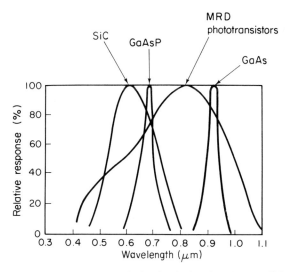

Figure 4-22 Spectral characteristics for GaAs, GaAsP, and SiC LEDs compared with MRD phototransistors (Courtesy Motorola)

MRD450 (which has a rated sensitivity of 0.8 mA/mW/cm²) when used with a GaAs LED is approximately: $0.8 \times 3.5 = 2.8$ mA/mW/cm².

Using LEDs with a lens. An additional factor in using LEDs is the polar response. The presence of a lens in the LED package will confine the solid angle of radiation, as is the case with any light source. When a lens is used with an LED, it is possible to predict the resultant irradiance H using the following equation

$$H = \frac{4P}{3.14 \times (\text{angle})^2 \times d^2} \text{ W/cm}^2$$

where:

P is the total output power of the LED in watts

(angle) is the solid angle in steradians

d is the distance between the LED and the detector in centimeters.

4-4. LOW-FREQUENCY AND STEADY-STATE DESIGN APPROACHES FOR PHOTOTRANSISTORS

The model shown in Fig. 4-23 is usually adequate for most low-frequency and steady-state circuit designs involving phototransistors. Note that any capacitance is omitted from the model. This is generally accurate for low-frequency or direct-current applications.

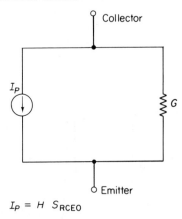

$I_P = H \ S_{RCEO}$

Figure 4-23 Low-frequency and steady-state model for floating-base phototransistor

Also note that the frequency range for which the model of Fig. 4-23 is valid depends on *load resistance*.

Figure 4-24 shows a plot of the 3 dB response frequency (in kilohertz) as a function of load resistance. Assume that a modulated light source is to drive the phototransistor at a maximum frequency of 10 kHz. If the resultant photo current is 100 μA, Fig. 4-24 shows a maximum load resistance of 8 kΩ for a frequency of 10 kHz. This means that the model of Fig. 4-23 can be used for loads of 8 kΩ, or less, if 100 μA is required at 10 kHz. If the load resistance is increased, and the frequency-current factors must remain the same, the model of Fig. 4-23 is not valid. The model of Fig. 4-3, or some similar model, must be used instead. Instead of concentrating on phototransistor models, let us apply the equations and design data to an actual low-frequency or steady-state problem.

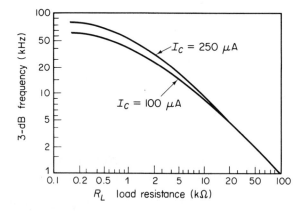

Figure 4-24 3-dB frequency versus load resistance for MRD phototransistors (Courtesy Motorola)

4-4.1. Light Operated Relay Using Phototransistor Control

Figure 4-25 shows a circuit in which the presence of light causes a relay to operate. When light is applied to Q_1, transistor Q_2 conducts and operates relay K_1. The relay used in this circuit draws about 5 mA when Q_2 is in saturation. The h_{fe} (minimum) for the MPS3394 (Q_2) is 55. A base current of 0.5 mA applied to Q_2 will therefore provide more than enough current to operate the relay when Q_2 saturates. In turn, phototransistor Q_1 will provide the 0.5 mA base current, if sufficient light is applied.

To find the required amount of light for Q_1, divide the required Q_2 base current by the illumination sensitivity of Q_1 [which is 4 μA/fc (fc is the abbreviation for footcandles.)] This results in a required illumination of: 0.5 mA/(4 $\times$ 10^{-3} mA/fc), or 125 fc. This light level can be supplied by a flashlight or any other equivalent light source. (About 200 fc can be expected from a flashlight with two D-cells.)

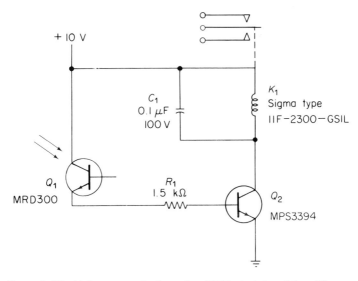

Figure 4-25 Light-operated relay using MRD phototransistors (Courtesy Motorola)

4-5. HIGH-FREQUENCY DESIGN APPROACHES FOR PHOTOTRANSISTORS

From a practical design standpoint, it is sufficient to consider a phototransistor as a *current source*, with a first-order transient response. With the addition of switching characteristics to the phototransistor information already discussed, most high-frequency design problems can be solved with a minimum of effort. Before going into some practical design examples, let us analyze the switching characteristics of typical phototransistors.

4-5.1. Phototransistor Switching Characteristics

When a phototransistor changes state from OFF to ON, a significant time delay is associated with the RC time constant of the base-emitter junction. As shown in Fig. 4-26, the capacitance of the base-emitter junction is large (over 60 pF for less than 1 V of forward bias). Since the device photocurrent is $g_m v_{be}$ (from Fig. 4-3), the load current can change state only as fast as v_{be} can change.

Also, v_{be} can change only as fast as C_{be} can charge and discharge through the load resistance. Figure 4-27 shows the variations in rise and fall time with load resistance. This measurement was made using a GaAs light emitting

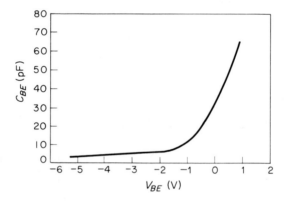

Figure 4-26 Base-emitter junction capacitance versus voltage for MRD300 phototransistor (Courtesy Motorola)

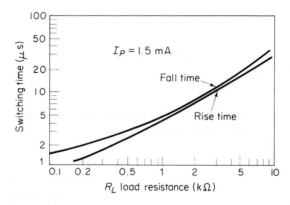

Figure 4-27 Switching time versus load resistance for MRD phototransistor (Courtesy Motorola)

diode for the light source. The LED output power, and separation distance between the LED and the phototransistor, were adjusted mechanically for an ON phototransistor current of 1.5 mA. The rise time was also measured for a short-circuited load and found to be about 700 nS.

The major difficulty found in high-frequency phototransistor applications is the *load-dependent frequency response.* Since the phototransistor is a current source, it is desirable to use a large load resistance to develop maximum output voltage. However, large load resistances limit the useful frequency range. This seems to present the designer with a tradeoff between voltage and speed. However, there is a technique available to eliminate the need for such a tradeoff.

Figure 4-28 shows a circuit designed to get the optimum speed and output voltage. The common-base stage Q_2 offers a low-impedance load to the phototransistor, thus producing best response speed. Since Q_2 has near-unity current gain (typical for grounded-base), the load current in R_L is approximately equal to the phototransistor current. Thus, the impedance transformation performed by Q_2 results in a relatively load-independent frequency response.

The effect of Q_2 is shown in Figs. 4-29 and 4-30. Figure 4-29 shows the

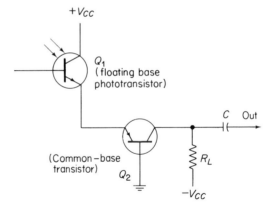

Figure 4-28 Circuit to get best speed and output voltage from phototransistor (Courtesy Motorola)

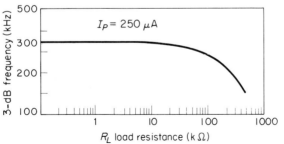

Figure 4-29 3-dB frequency versus load resistance for MRD phototransistors using common-base speed-up circuit (Courtesy Motorola)

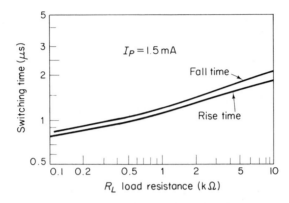

Figure 4-30 Switching time versus load resistance for MRD300 phototransistor using common-base speed-up circuit (Courtesy Motorola)

3-dB frequency response as a function of load. Comparing Fig. 4-29 with Fig. 4-24, shows that the effect of Q_2 is quite pronounced. Comparing Fig. 4-30 with Fig. 4-27 also demonstrates the effect of Q_2.

4-5.2. Using Phototransistors in Logic Circuits

One of the most common uses for phototransistors is in logic circuits. It is possible to provide logic control of power devices with optical signals. The circuits of Figs. 4-31 through 4-34 are typical of those used to convert light signals into electrical signals. The truth tables for these circuits show outputs for positive logic.

For the circuit shown in Fig. 4-31, an output of 1 is obtained at all times except when both Q_1 and Q_2 are exposed to bright light. Any light source with approximately 100 fc to Q_1 and Q_2 will drive Q_3 into saturation. Resistor R_2 provides a path for leakage currents so that Q_3 does not conduct until an adequate light level is presented to Q_1 and Q_2. The positive output will provide almost 2 mA to a load.

The circuit shown in Fig. 4-32 provides a positive output voltage at all times, except when Q_1 is on and Q_2 is off. For this case, the base drive to Q_3 is provided through R_1 and Q_1. Normal room lighting will not turn Q_3 on, but about 100 fc from a flashlight shining on Q_1 will allow Q_3 to saturate. The maximum current from the Fig. 4-32 circuit to a load is 2 mA.

The inverse of the circuits shown in Figs. 4-31 and 4-32 can be obtained by means of an additional inverting stage. An alternate procedure is to use the circuits of Figs. 4-33 and 4-34. If a current sink is required, use the additional inverter stage. If a current source is required, use the circuits of Figs. 4-33 and 4-34.

Sec. 4-5   High-Frequency Design Approaches for Phototransistors   103

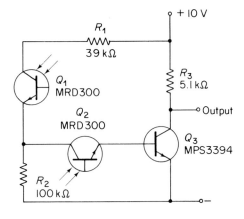

Figure 4-31 Optical logic driver where output is "0" only when Q_1 and Q_2 are ON (Courtesy Motorola)

Truth Table

Q_1, Q_2	Output = 0
$Q_1, \bar{Q}_2$	Output = 1
$\bar{Q}_1, Q_2$	Output = 1
$\bar{Q}_1, \bar{Q}_2$	Output = 1

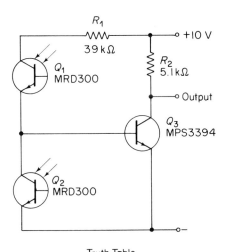

Figure 4-32 Optical logic driver where output is "0" only where Q_1 is ON and Q_2 is OFF (Courtesy Motorola)

Truth Table

$Q_1, \bar{Q}_2$	Output = 0
Q_1, Q_2	Output = 1
$\bar{Q}_1, \bar{Q}_2$	Output = 1
$\bar{Q}_1, Q_2$	Output = 1

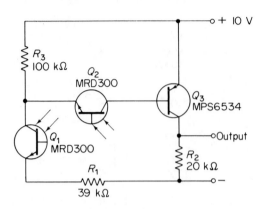

Figure 4-33 Optical logic driver where output is "1" only where Q_1 and Q_2 are ON (Courtesy Motorola)

Truth Table

Q_1, Q_2	Output = 1
$Q_1, \bar{Q}_2$	Output = 0
$\bar{Q}_1, \bar{Q}_2$	Output = 0
$\bar{Q}_1, Q_2$	Output = 0

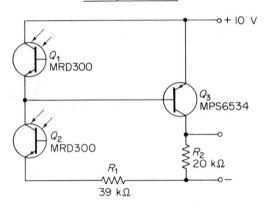

Figure 4-34 Optical logic driver where output is "1" only where Q_1 is OFF and Q_2 is ON (Courtesy Motorola)

Truth Table

$\bar{Q}_1, Q_2$	Output = 1
Q_1, Q_2	Output = 0
$\bar{Q}_1, \bar{Q}_2$	Output = 0
$Q_1, \bar{Q}_2$	Output = 0

The output of the Fig. 4-33 circuit is 0 at all times, except when both Q_1 and Q_2 are on. As in the previous circuits, about 100 fc of illumination on Q_1 and Q_2 is enough to saturate Q_3 (and permit Q_3 to supply about 10 mA to a load). The value of R_2 is high so that it draws very little current away from the load. A current of about 0.5 mA in R_2 is typical, and produces satisfactory results. Resistor R_3 provides a path for leakage currents so that Q_3 does not conduct until Q_1 and Q_2 are illuminated.

The circuit of Fig. 4-34 provides a zero output for all conditions except when Q_1 is off and Q_2 is on. For this condition, base current is provided to Q_3 through Q_2 and R_1. Here also, 100 fc is enough to allow Q_3 to saturate, and drive a 10-mA load. For the Fig. 4-34 circuit, the value of R_2 is also high so that R_2 does not rob current from the load. Q_1 must be turned on hard enough to shunt the base-emitter junction of Q_3 to keep Q_3 from coming on when Q_2 is on. The 100-fc illumination is sufficient for this purpose.

4-5.3. Using LEDs with Phototransistor Logic Circuits

It is quite common to use LEDs as the light source for phototransistor logic circuits. Such an arrangement is particularly useful when it is desired to maintain high isolation between signal and power circuitry. Often, both the LED and phototransistor are housed in a common enclosure so that the light path is unobstructed (and that the path is shielded from other light sources). Commercially-available devices that use such an arrangement are called *photocouplers* or simply *couplers*.

The main problem in designing a photocoupler is to find the correct spacing between the LED and the phototransistor. This can be done by rearranging the equation of Sec. 4-3.5. The following is an example of how the equation can be rearranged to find the correct distance between an LED and the phototransistors (MRD300) Q_1 and Q_2 of Fig. 4-31.

Assume that the LED in question has an output power (P) of 1.5 mW (minimum) at an I_D of 100 mA, a beam angle of 20° and a spectral response as shown in Fig. 4-22. As previously discussed, an LED of the type shown in Fig. 4-22 has an efficiency of about 90 per cent when used with MRD series phototransistors. Thus, the equivalent power output of the LED is: 1.5 mW $\times$ 0.9 = 1.35 mW.

Also, as discussed in Sec. 4-4.1, the MPS3394 Q_3 will saturate with a base current of about 0.5 mA. Thus, the LED must provide sufficient light so that Q_1 and Q_2 will provide at least 0.5 mA photocurrent to Q_3.

From the MRD300 datasheet, S_{RECO} (collector-emitter radiation sensitivity) is found to be about 0.55 mA/mW/cm² minimum at $H = 1.0$ mW/cm². That is, for an H of 1, the photocurrent output of Q_1 or Q_2 is about 0.55 mA, and Q_3 will saturate.

To find the correct distance at which the LED will produce the required light, rearrange the equation of Sec. 4-3.5. as follows

$$d = \sqrt{\frac{4P}{H \times 3.14 \times (\text{angle})^2}}$$

where:
    d is the distance in centimeters
    H is the required irradiance in milliwatt/centimeters
    P is the equivalent power output of the LED in milliwatts
    (angle) is the LED beam angle in degrees.

Note that the equation must be solved in terms of radians rather than degrees. One degree equals 0.01745 rad, and $20° = 0.349$ rad.

Using the established values in this equation, the distance is found to be

$$d = \sqrt{\frac{4 \times 1.35}{1 \times 3.14 \times (0.349)^2}} \approx \sqrt{14} \approx 3.75 \text{ cm}$$

Thus, if the LEDs are axially aligned with the MRD300s (Q_1 and Q_2) at a distance *within* 3.75 cm, saturation of Q_3 is assured.

5 AF AMPLIFIER DESIGN EXAMPLES

Both two-junction transistors and FETs are used as AF amplifiers. UJTs are not generally found in AF amplifier work. One exception is the UJT regenerative amplifier discussed in Sec. 5-4. FETs are usually limited to voltage amplifier applications. Two-junction transistors can be used in either voltage amplifier or power amplifier circuits. Since the number of different amplifier circuits is almost unlimited, it is impossible to cover all aspects of AF amplifier circuit design in this chapter. Instead, we shall concentrate on a *cross section* of AF amplifier design examples. The author's *Handbook of Modern Solid-State Amplifiers* (Prentice-Hall, Inc., Englewood Cliffs, N.J., 1974) provides a detailed discussion of many amplifier types.

All basic design considerations for transistors apply to AF amplifiers. Of particular importance are: how to interpret datasheets, determining parameters at different frequencies, temperature-related design problems, and basic bias schemes. All of these subjects are discussed in the author's *Handbook for Transistors* (Prentice-Hall, Inc., Englewood Cliffs, N.J., 1976).

5-1. THE EFFECT OF AMPLIFIER COMPONENTS ON FREQUENCY

Amplifier components do not attenuate (or pass) signals of all frequencies equally. That is, every electronic component has some *impedance* and is thus frequency *sensitive*. Even a simple length of wire has some impedance. Wire, being a conductor, has some resistance. If alternating current is passed through the wire, there is some inductive reactance. If the wire is near another conductor (or metal chassis), there is some capacitance between the two conductors, and thus some capacitive reactance. The reactance and resistance combine to produce impedance, which, in turn, varies with frequency.

In theory, transistors are capable of operating at any frequency from direct-current on up. The top frequency limit is set only by the transit time of electrons across the transistor junctions. Primarily because of reactance, there are limitations placed on the operating frequency of any amplifier by the transistor characteristics. Also, other components in the amplifier (resistors, capacitors, inductances, etc.) limit the operating frequency of any circuit. In this section, we shall see how each of the components affects operating frequency of amplifier circuits from a design standpoint.

In practical design work, many of the impedances and reactances presented by components are of little concern. In other cases, certain impedances and reactances have a very pronounced effect on amplifier design. In addition to transistors, other major components used in AF amplifier design include capacitors, resistors, and inductances (both coils and transformers). Let us see how the impedances and reactances of these components affect AF amplifier operation.

Transistor frequency limitations. All transistors have some capacitance between the elements (emitter-base junction, collector-base junction, gate-drain, etc.). If any of the elements is common or ground, the remaining elements have some capacitance to ground. For example, in a common-source amplifier, there is some capacitance from gate to ground (across the input) and drain to ground (across the output). Likewise, there is capacitance from the drain to gate (which forms a feedback path from output to input). This is shown in Fig. 5-1.

Capacitive reactance decreases with an increase in frequency, and vice

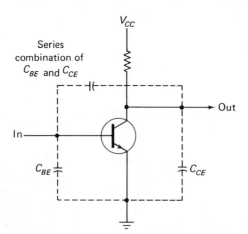

Figure 5-1 Capacitances associated with transistor elements

versa. A capacitance in series with a conductor presents less attenuation to the signal as frequency increases. A capacitance across a conductor (for example, in parallel or shunt from the conductor to ground) acts as a short to signals of increasing frequency. Consider a common-source amplifier where FET capacitances are across the input and output. As frequency increases, the capacitive reactance drops, producing a short across the input and output, and increased attenuation of the signal. At some frequency, the attenuation equals the FET amplification, so there is no gain. At higher frequencies, the attenuation exceeds amplification, and there is a loss, even though the FET may still continue to operate.

From a circuit design standpoint, the input and output capacitances of transistors have little effect at audio frequencies. Most modern transistors will operate well beyond the AF range, and will generally produce equal (or flat) frequency response. That is, all signals up to about 20 kHz (or possibly higher) are amplified by the same amount. However, with most transistors, as signal frequencies increase into the RF range, amplification begins to drop. (RF amplifier circuit design is discussed in Chapter 6.)

All transistors have some inductance in their leads. This produces inductive reactance in series with the transistor elements. Inductive reactance increases with frequency. In the AF range, the inductive reactance is of little concern. However, at radio frequencies, the inductive reactance can produce considerable attenuation.

Resistor frequency limitations. At audio frequencies, resistors offer relatively few problems, since resistors attenuate signals equally. Only at very high frequencies, where the resistor leads and body could produce some kind of reactance, is there any particular concern about frequency limits imposed by resistors. However, resistors do produce *voltage drops*. These voltage drops can be a problem when considering interstage coupling methods (described in Sec. 5-2). Also, resistors offer some problems when used in conjunction with coupling capacitors in amplifier circuits.

Capacitance frequency limitations. Capacitors have three major uses in transistor AF amplifier circuits: bypass, decoupling, and coupling.

Bypass capacitors are used to provide a signal path around high resistances. For example, if the power supply of an AF amplifier does not have a filter capacitor, or a battery is used, the collector-emitter current must pass through a high resistance. This can impede the a-c component of the signal. A bypass capacitor provides a signal path as shown in Fig. 5-2.

When several stages of amplification are connected, they all join at one point, the common power supply. In multistage amplifiers there is the possibility of one stage feeding back through the power supply to a previous stage, thus causing interference with the signal. To avoid this feedback, one or more of these stages may be *decoupled* from the power supply. Figure 5-2

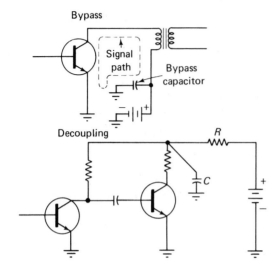

Figure 5-2 Examples of bypass and decoupling capacitors

also shows a typical decoupling network. Resistor R is placed in series between the load resistors of the stages and the power supply. Hence, resistor R offers a high-resistance path from the a-c signal component to the power supply. Capacitor C, conversely, offers a low shunt reactance to this component, and thus decouples (bypasses) the component to ground.

Actually, the functions of bypass and decoupling capacitors are the same, and the terms are interchangeable. In either case, the main concern is that the *reactance be low at the lowest frequency involved*. For example, assume that the lowest frequency is 100 Hz, and the minimum required reactance is 100 Ω. This requires a reactance of about 16 μF ($C = 1/(6.28F\, X_c)$). If the required frequency is decreased to 10 Hz, the capacitance value must be raised to about 160 μF to keep the reactance below 100 Ω.

Coupling capacitors are used at the input and output of each stage to block direct current. For example, if a coupling capacitor is not used between the transistor stages, the collector of the first stage is connected directly to the base of the following stage, and both elements are at the same bias voltage. While it is possible to operate transistors in this way, direct coupling does increase problems. For one thing, the amplifier cannot tell the difference between a change of signal level and a change in power supply voltage.

The values of coupling capacitors are dependent upon the *low-frequency limit* at which the amplifier is to operate, and on the resistance with which the capacitors operate. As frequency increases, capacitive reactance decreases and coupling capacitors become (in effect) a short to the signal. Therefore, the high-frequency limit need not be considered in AF circuits.

Figure 5-3 shows how a high-pass filter is formed by coupling capacitors. Capacitor C_1 forms a high-pass RC filter with R_B. Capacitor C_2 forms another

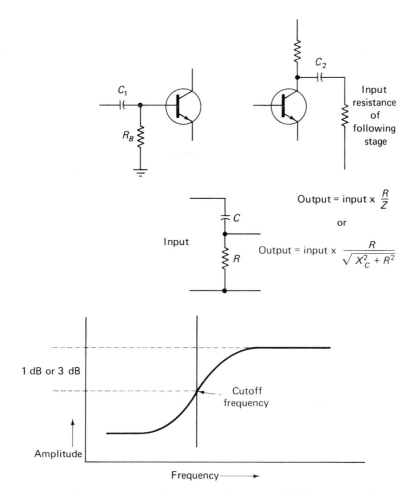

Figure 5-3 Formation of high-pass (low-cut) *RC* filter by coupling capacitors and related resistances

high-pass filter with the input resistance of the following stage (or the load). The input voltage is applied across the capacitor and resistor in series. The output voltage is taken across the resistance. The relationship of input voltage to output voltage is

$$\text{Output voltage} = \text{input voltage} \times \frac{R}{Z}$$

where R is the d-c resistance value, and Z is the impedance obtained by the vector combination of series capacitive reactance and d-c resistance.

As a design guideline, when reactance drops to about one-half the resistance, the output will drop to about 90 per cent of the input (or approximately a 1-dB loss). Using the 1-dB loss as the low-frequency cutoff point, the value of C_1 or C_2 can be found by

$$\text{Capacitance} = \frac{1}{3.2FR}$$

where capacitance is in farads, F is the low-frequency limit in hertz, and R is resistance in ohms.

If a 3-dB loss can be tolerated at the low-frequency cutoff point, the value of C_1 or C_2 can be found by

$$\text{Capacitance} = \frac{1}{6.2FR}$$

Inductance frequency limits. Both coils and transformers are used in audio amplifiers. As discussed in Sec. 5-2, coils are sometimes used in place of the collector resistor as a load. This permits the collector to be operated at the full voltage. Likewise, transformers are used for coupling between stages. This provides impedance matching, as discussed in Sec. 5-2.

The inductive reactance of coils and transformers increases with frequency. At the high end of the audio range, the attenuation produced by this increased reactance is usually sufficient to impair operation of the amplifier. At the low end of the AF range, the reactance of a typical transformer drops to a few ohms. This low impedance acts as a short across the line and attenuates the signal. Thus, coils and transformers tend to attenuate signals at both the high end and low end of the AF range.

Stray impedances. As discussed, any conductor (wiring, terminals, etc.) can have resistance, reactance, and impedance. Thus, care must be used in the routing of wires and placement of terminals to minimize the effects of this *stray* impedance. Likewise, the effects of stray impedance can alter the characteristics of components. A classic example of this is stray capacitance, which is added to the input and output capacitances of transistors. The effects of stray impedances are usually not critical at audio frequencies. However, the effects of stray impedance on amplifiers operating at radio frequencies (Chapter 6) can be of considerable importance.

5-2. AMPLIFIER COUPLING CIRCUIT DESIGN

The design of coupling networks must be considered in any amplifier circuit. Even a single-stage audio amplifier must be coupled to the input and output devices. If the circuit is multistage, there must be interstage coupling. Amplifier circuits are often classified according to coupling design.

For example, the four basic coupling design techniques are: capacitor (or capacitance), inductive, direct, and transformer.

The term *resistance-coupled* could be applied to any of the four coupling design techniques. However, the term resistance-coupled is generally used to indicate that the circuit does not have inductances or transformers between stages, and that the input and/or output impedance is formed by a resistance. Capacitor coupling is often called *resistance-capacitance* (or *RC*) coupling.

In this section, we shall see how the different design techniques affect operation of solid-state AF amplifiers. Figure 5-4 shows the four coupling methods.

(1) With direct coupling, Fig. 5-4(a), the collector of one transistor is connected directly to the base of the following transistor. The outstanding characteristic of a direct-coupled amplifier is the ability to amplify direct current and low-frequency signals.

(2) With capacitor coupling, or RC coupling, Fig. 5-4(b), the coupling is

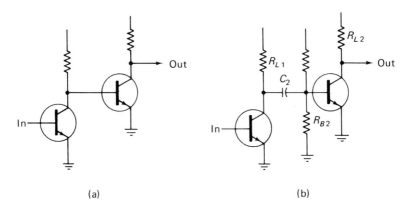

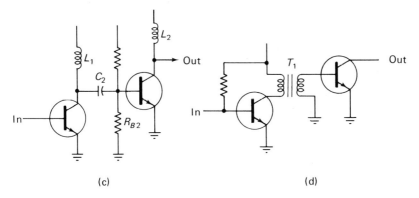

Figure 5-4 Four basic types of coupling used in audio amplifiers

accomplished by means of the load resistor R_{L1} of stage 1, the base resistor R_{B2} of stage 2, and the coupling capacitor C_2. The original signal is acted upon by stage 1 and appears in amplified form as the voltage drop across R_{L1}. The d-c component of the amplified signal is blocked by C_2 which passes the a-c component to the input section of stage 2 for further amplification. If necessary, more stages may be coupled to the output of stage 2 for still further amplification.

The main advantage of capacitor or *RC* coupling is that the amplifier will amplify uniformly over nearly the entire audio range, since resistor values are independent of frequency changes. However, as dicussed in Sec. 5-1, *RC* coupling amplifiers do have a low-frequency limit imposed by reactance of the capacitor (which increases as frequency decreases). Reactance-capacitance coupling is also small, light, inexpensive, and produces no magnetic field to interfere with the signal. One disadvantage of the *RC* coupling method is that the supply voltage is dropped (usually to one-half) by the load resistance. Thus, the collectors must operate at reduced voltages.

(*3*) *With inductive or impedance coupling*, Fig. 5-4(c), the load resistors R_{L1} and R_{L2} are replaced by inductors L_1 and L_2. The advantage of impedance coupling over resistance coupling is that the ohmic resistance of the load inductor is less than that of the load resistor. Thus, for a power supply of a given voltage, there is a higher collector voltage.

Impedance coupling also suffers from some disadvantages. Impedance coupling is larger, heavier, and more costly than resistance coupling. To prevent the magnetic field of the inductor from affecting the signal, the inductor turns are wound on a closed, iron core and usually shielded extensively. The main disadvantage of impedance coupling is frequency discrimination.

With impedance coupling at very low frequencies, the gain is low due to the capacitive reactance of the coupling capacitor, just as in the *RC*-coupled amplifier. The gain increases with frequency, leveling off at the middle frequencies of the audio range. (However, the frequency spread of this level portion is not as great as for the *RC* amplifier.)

With impedance coupling at very high frequencies, the gain drops because of the increased reactance. Impedance coupling is rarely, if ever, used at frequencies above the audio range.

(*4*) *With transformer coupling*, Fig. 5-4(d), the transformer T_1 serves several purposes. As the fluctuating collector current of the first stage flows through the primary winding of T_1, the current induces an alternating voltage with similar waveform in the secondary of T_1. This voltage forms the input signal to the second stage. Since the secondary of T_1 conveys the a-c component of the signal directly to the base of the second stage, there is no need for a coupling capacitor. Also, since the secondary winding furnishes a base return path, there is no need for a base resistance.

Compared to the *RC*-coupled amplifier, the transformer-coupled amplifier has essentially the same advantages and disadvantages as the impedance-coupled amplifier. The transistor collectors can be operated at higher voltages. The impedances are set by the transformer primary and secondary windings. Transformers are frequency sensitive (impedance changes with frequency). Therefore, the frequency range of the transformer-coupled amplifiers is limited.

The inductances and transformers used in AF work are generally of the iron-core type. If air-core transformers are used at audio frequencies, the inductive reactance (and the impedance) will be so small as to be ineffective. At frequencies above the audio range (or at the high end), the reactance of iron-core inductances and transformers is so large that signals cannot pass (or are greatly attenuated). Thus, air-core transformers and inductances are used for higher-frequency amplifiers, as discussed in Chapter 6.

Coupling transformers also provide for impedance matching between stages. Because the transistor is a current-operated device, impedance matching between the output of one stage to the input of the next is desirable for maximum transfer of power. This can be accomplished by making the primary and secondary transformer winding of different impedances.

Typically, the input impedance of a transistor stage is less than the output impedances. Thus, the secondary impedance of an interstage transformer is typically lower than the primary impedance. When two common-emitter stages are impedance matched, the overall gain is greater than when identical stages are resistance coupled.

Transformer coupling is also effective when the final amplifier output must be fed to a low-impedance load. For example, the impedance of a typical loudspeaker is in the order of 4 to 16 Ω, whereas the output impedance of a transistor stage is several hundred (or thousand) ohms. A transformer at the output of an audio amplifier can offset the obviously undesired effects of such a mismatch.

5-2.1. Effects of Coupling on Audio-Amplifier Frequency Response

A simplified *frequency-response graph* or curve is illustrated in Fig. 5-5, which shows the effects of coupling methods on amplifier frequency response. The response is measured by the gain of the amplifier at various frequencies in the audio range.

Note that the gain falls off at the very low frequencies. In an *RC*-coupled amplifier, this drop in gain (generally referred to as *rolloff*) at the low end is due to the capacitive reactance of the coupling capacitor. Since the coupling capacitor is between the output of the first stage and the input to the second stage, the signal is attenuated by the voltage drop across the capacitor. Hence, the lower the frequency, the larger the capacitive reactance, and the smaller

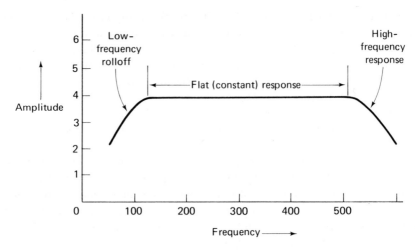

Figure 5-5 Simplified frequency-response graph

the signal input is to the second stage. In impedance-coupled or transformer-coupled amplifiers, the low-frequency rolloff is caused by the very low inductive reactance, which acts as a short across the signal path. In effect, the low reactance bypasses some of the signal to ground.

As shown in Fig. 5-5, the gain also falls off at the higher frequencies. In *RC*-coupled amplifiers, this high-frequency rolloff is due to the output capacitance of the first stage, the input capacitance of the second stage, and the stray capacitance furnished by the coupling network. These capacitances act to bypass some of the signal to ground. The higher the frequency, the smaller the capacitive reactance becomes, and the greater the amount of signal so bypassed. Hence, the overall gain falls. At frequencies between these two extremes, the gain remains fairly constant. In impedance-coupled or transformer-coupled amplifiers, the high-frequency rolloff is due to the large inductive reactance that attenuates the signal.

In sum, then, resistance coupling produces the lowest gain, transformer coupling the highest. As a guideline, three stages of *RC*-coupled amplification will produce approximately the same gain as two stages of *comparable* transformer-coupled amplification. *RC*-coupling produces the least frequency distortion of the signal. Transformer coupling has the added advantage of providing an impedance match at the amplifier input and output.

5-3. AMPLIFIER DESIGN CLASSIFICATIONS

There are many classification methods for amplifier design. One of the most common is to classify amplifiers by *operating point*. That is, the particular design is classified by the amount of current flowing in the amplifier transistors under no-signal conditions. The following is a brief summary of the four basic operating-point classifications.

In all four classifications, the base-collector (or gate-drain) junction is always reverse-biased at the operating point, as well as under all signal conditions. Thus, no base-collector current flows (with the possible exception of reverse leakage current I_{CBO}). The base-emitter (or gate-source) junction is biased such that base-emitter will flow under certain conditions, and possibly under all conditions. When base-emitter current flows, emitter-collector current also flows.

5-3.1. Class A Amplifier

In the class A amplifier, the base-emitter bias and the input voltages are such that the transistor operates only over the *linear portion* of the *characteristic curve*. Such a curve, representing the relationship between base voltage (input) and collector current (output), is shown in Fig. 5-6. At no point of the input signal cycle does the base become so positive or negative as to cause the transistor to operate at the nonlinear portion of the curve. The transistor collector current is never cut off, nor does the transistor ever reach saturation.

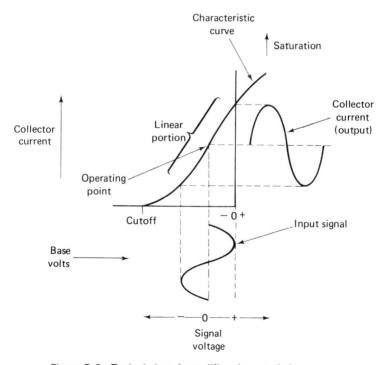

Figure 5-6 Typical class A amplifier characteristic curves

The main advantage of the class A amplifier is the relative lack of distortion. The output waveform follows that of the input waveform, except in amplified form. However, with any class of amplifier there is some distortion.

The disadvantages of class A amplifiers are their relative inefficiency (low power output for a high power input dissipated by the transistor), and their inability to handle large signal voltage swings. Rarely is a class A amplifier over about 35 per cent efficient. Thus, if the power input to a class A amplifier is 1 W (generally, the maximum power dissipation capability of a single transistor), the output will be less than 0.3 W.

The peak-to-peak output signal voltage swing of a class A amplifier is limited to something less than the total supply voltage. Since the output voltage must swing both positive and negative, the peak output is less than one-half the supply voltage. For example, assume that the supply voltage is 20 V and the amplifier is biased so that the Q-point collector voltage is one-half the supply, or 10 V. (Such a Q point is generally typical for a class A amplifier.) Under these conditions, the output voltage swing cannot exceed ± 10 V. If distortion must be kept to a minimum, the output will usually be on the order of ± 5 V, so as to keep the transistor on the linear portion of the characteristic curve. (In most cases, the curve becomes nonlinear near the cutoff and saturation points) However, this can be determined only from an actual test of the amplifier circuit.

The input voltage swing of a class A amplifier is limited by the output voltage swing capability and the voltage amplification factor. For example, if the output is limited to ± 10 V, and the voltage amplification factor is 100, the input is limited to ± 0.1 V (100 mV).

Because of these limitations, class A amplifiers are generally used as voltage amplifiers, rather than power amplifiers. Typically, a class A amplifier stage is used ahead of a power amplifier stage.

5-3.2. Class B *Amplifier*

If the base-emitter bias is changed so that the operating point coincides with the transistor cutoff point, we obtain class B amplification. For an NPN transistor this means making the base more negative than for class A operation. (For PNP transistors, class B is obtained by making the base more positive than for class A.) Either way, the base-emitter reverse bias is increased (or forward bias is decreased) for class B operation.

As shown in Fig. 5-7, when the input signal voltage is zero, there is no flow of collector current. During the positive half-cycle of the signal voltage (Fig. 5-7 is for an NPN transistor), the collector current rises to its peak and then falls back to zero in step with the variations of that half-cycle. During the negative half-cycle of the signal-voltage, there is no collector current since the

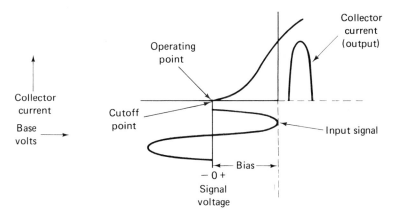

Figure 5-7 Typical class B amplifier characteristic curves

base-emitter reverse bias is at all times greater than the cutoff voltage of the transistor. Hence, collector current flows only during half the input signal cycle.

If a single transistor is operated as class B, there will be considerable distortion. This is because the waveform of the resulting collector current resembles that of the positive half-cycle of the input signal and, consequently, does not resemble the complete waveform of the input. It is possible to use two transistors, one for each half-cycle of the input signal, and by combining the outputs of these transistors in *push-pull*, to reconstruct an output whose waveform resembles the full waveform of the input.

The peak output voltage swing of a class B amplifier is slightly less than the supply voltage. Since the output appears only on half-cycles, it is possible to operate class B amplifiers at a higher current (or power) rating than class A, all other factors being equal. For example, if a transistor is capable of 0.3-W dissipation (without damage) as class A, the same transistor can be operated at 0.6 W, class B, since the transistor is conducting collector current only half the time. (This is a theoretical example. In practice, there are factors which limit class B power dissipation to something less than twice that of class A.)

The peak output of a class B amplifier is equivalent to the peak-to-peak output of a class A amplifier. Thus, if two transistors are connected in push-pull and operated as class B, the *output voltage can be twice* that of class A.

Because of these voltage and power factors, class B amplifiers are generally used as power amplifiers, rather than voltage amplifiers. Typically, two push-pull transistors are operated in class B, preceded by a single class A amplifier stage. The class A stage provides voltage amplification, whereas the class B stage produces the necessary power amplification.

5-3.3. Class AB Amplifier

Class B is the most efficient operating mode for audio amplifiers, since it draws the least amount of current. That is, the transformers are cut off at the Q point and draw collector current only in the presence of an input signal. Class B operation can, however, result in a form of distortion known as *crossover distortion*. The effects of crossover distortion can be seen by comparing the input and output waveforms on Fig. 5-8.

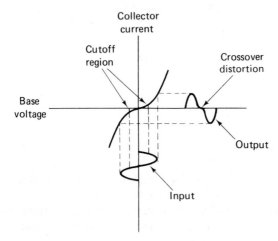

Figure 5-8 Example of crossover distortion

In true class B operation, the transistor *remains cut off* at very low-signal inputs (because transistors have very low current gain at cutoff) and turns on abruptly with a large signal. For example, a silicon transistor does not have appreciable collector current flow until the base-emitter junction is forward-biased by about 0.5 V. Assuming that the input signal starts at 0 V, there will be little or no collector current flow (and thus no change in the output voltage) during the time that the input signal is going from 0 to 0.5 V. When the input reaches 0.5 V, the collector current increases rapidly and follows the input signal in a linear fashion.

Crossover distortion can be minimized by operating the stage as class AB (or somewhere between B and AB). That is, the transistors are forward-biased just enough for a *small amount* of collector current to flow at the Q point. For a typical silicon transistor, the forward bias is just below 0.5 V for class AB. Thus, some collector current is flowing at the lowest signal levels, and there is no abrupt change in current gain. Class AB is less efficient than class B because more current must be used. Generally, class AB is only used in push-pull circuits.

5-3.4. Class C Amplifier

If the transistor is reverse-biased considerably below the cutoff point, we obtain a class C amplifier. As shown in Fig. 5-9, during the positive half-cycle of the input signal, the signal voltage starts from zero, rises to the positive peak value, and falls back to zero. (Fig. 5-9 is for an NPN transistor.) A *portion* of the input signal causes the base-emitter junction to be forward-biased. As a result, there is a flow of collector current for a portion of one-half the input cycle. The negative half-cycle of the input signal lies well below the cutoff point of the transistor. Collector current flows only during that portion of the positive half-cycle of the input signal between the cutoff point and the peak. The resulting collector current is a pulse, the duration of which is considerably less than a half-cycle of the input signal.

Obviously, the waveform of the output signal cannot resemble that of the input signal. Nor can this resemblance be restored by the push-pull method mentioned in the discussion of class B and AB amplifiers. For this reason, class C is limited to those applications where distortion is of no concern. Generally, class C operation is limited to use in RF amplifiers (Chapter 6), and is not found in AF amplifiers.

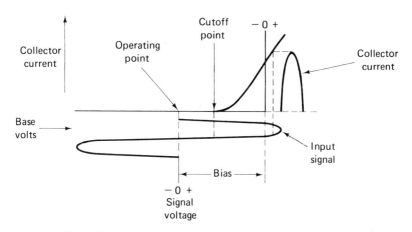

Figure 5-9 Typical class C amplifier characteristic curve

5-4. UJT REGENERATIVE AMPLIFIER

The basic UJT relaxation oscillator described in Chapter 8 can be adapted to form a regenerative pulse amplifier. Such an amplifier is shown in Fig. 5-10. Note that the circuit is essentially the basic relaxation oscillator with two resistors added. Resistor R_L provides a load for a trigger input or a pulse output. Resistor R_3 forms a voltage divider with resistor R_E. The

122 AF Amplifier Design Examples Chap. 5

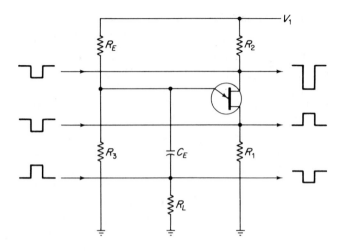

$R_L \approx Z$ OF TRIGGER, OR Z OF OUTPUT LOAD

$$R_3 \approx \frac{R_E}{V_1 - V_{BIAS}} \times V_{BIAS}$$

$$R_E \gg \frac{V_1 - V_V}{I_V} \qquad V_{BIAS} = V_P - 0.5 \text{ TRIGGER}$$

$$C_E \approx \frac{\text{PERIOD}}{R_E \times \ln \frac{1}{1-\text{STANDOFF RATIO}}}$$

Figure 5-10 Basic UJT regenerative amplifier (Courtesy General Electric)

voltage at the junction of R_3 and R_E is set so that the UJT will not fire except in the presence of a trigger. That is, R_3 and R_E have a ratio such that the emitter voltage does not exceed the peak-point (firing) voltage for the off-state.

As shown in Fig. 5-10, inputs can be applied to any of three points (Base 1, Base 2, or emitter). Likewise, outputs can be taken from any of these points. In operation, a trigger pulse is applied to one of the inputs, and a corresponding trigger output (in greatly amplified form) is taken from another of the points. For example, a trigger of 0.1 V can be applied to the emitter, with a 7-V output taken from Base 1.

5-4.1. Design Considerations

The values of R_1, R_2 and C_E are selected on the same basis as discussed for the UJT relaxation oscillator in Chapter 8. The time constant for R_E and C_E should be approximately the same as the period of the trigger.

The value of R_E is found in the same way essentially as for the basic oscillator, but with a minor difference. The maximum value of R_E need not be of great concern in the regenerative amplifier. The UJT is held in the OFF condition until a trigger is applied by the voltage drop across R_3 (or by the voltage division at the junction of R_E and R_3). Thus, no calculating need be made for $R_{E(\max)}$. The minimum value for R_E is found in the same way as for the basic oscillator. That is, the minimum R_E that can be used with the regenerative amplifier is set by

$$R_E > \frac{V_1 - V_V}{I_V} = R_{E(\min)}$$

In order to assure turn-off after the trigger pulse is removed, however, the value of R_E should be two to three times larger than $R_{E(\min)}$.

The value of R_L is selected to match the load impedance of the input trigger source, or the output, depending upon application.

The value of R_3 is set by the desired bias voltage. The ratio of R_E and R_3 sets the fixed bias that must be overcome by the input trigger. V_{bias} is set at V_P (peak voltage of the UJT), less one-half the input trigger.

5-5. BASIC TWO-JUNCTION TRANSISTOR AMPLIFIER STAGE

Figure 5-11 is the working schematic of a basic, single-stage audio amplifier using a two-junction transistor. Input and output coupling capacitors C_1 and C_2 are added to prevent d-c flow to and from external circuits. A bypass capacitor C_3 is shown connected across the emitter resistor R_E. Capacitor C_3 is required only under certain conditions, as discussed in later paragraphs.

Input to the amplifier is applied between base and ground across R_B. Output is taken across the collector and ground. The input signal adds to, or subtracts from, the bias across R_B. Variations in bias voltage cause corresponding variations in base current, collector current, and the drop across collector resistor R_L. Therefore, the collector voltage (or circuit output) follows the input signal waveform, except that the output is inverted in phase. (If the input swings positive, the output swings negative, and vice versa.)

Variations in collector current also cause variations in emitter current. This results in a change of voltage drop across the emitter resistor R_E and a change in the base-emitter bias relationship. The change in bias that results from the voltage drop across R_E tends to cancel the initial bias change caused by the input signal, and serves as a form of negative feedback to increase stability (and limit gain). This form of emitter feedback (current feedback) is known as *stage feedback* or *local feedback*, since only one stage is

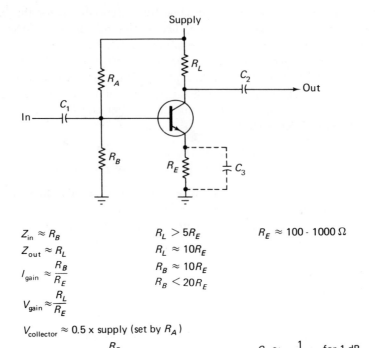

Figure 5-11 Basic single-stage audio amplifier using a two-junction transistor

involved. As discussed in later sections, *overall feedback* or *loop feedback* is sometimes used where several stages are involved.

5-5.1. Circuit Analysis

The outstanding characteristic of the circuit in Fig. 5-11 is that circuit characteristics (gain, stability, impedance) are determined (primarily) by circuit values, rather than transistor characteristics (β). The circuit is shown with an NPN transistor. The power supply polarity must be reversed if a PNP transistor is used.

The maximum peak-to-peak output voltage is set by the source voltage. For class A operation, the collector is operated at approximately one-half the source voltage. This permits the maximum positive and negative swing of output voltage. Generally, the absolute maximum peak-to-peak output can be between 90 and 95 per cent of the supply. For example, if the supply is 20 V, the collector will operate at 10 V (Q point), and swing from about 1 to 19 V. However, there is less distortion if the output is one-half to one-third of the source. In any circuit, the maximum collector voltage rating of the transistor cannot be exceeded.

The input and/or output impedances are set by the values of R_B and R_L, as shown by the equations of Fig. 5-11. Maximum power transfer occurs when R_B and R_L match the impedances of the previous stage and following stage, respectively.

Stability versus gain trade off. As shown by the equations, maximum voltage gain occurs when R_L is made larger in relation to R_E. Likewise, current gain increases when R_B is made larger in relation to R_E. However, circuit stability is greatest when R_B and R_L are made smaller in relation to R_E. That is, the circuit gain will remain more constant in the presence of temperature, supply voltage, or input signal changes when R_E is made larger in relation to R_B and R_L. Thus, there is a tradeoff between gain and stability.

In a practical amplifier, the circuit of Fig. 5-11 should be limited to a maximum current gain of 10 and a maximum voltage gain of 20. Higher gains are possible, but the stability is generally poor. Of course, even though gain is set by circuit values, the minimum a-c β of the transistor must be higher than the desired gain. For example, if the circuit values are chosen for a gain of 20, the minimum β must be 20 across the entire frequency range.

The Q point is affected by the values of all four resistors in Fig. 5-11. However, the value of R_A is the *final determining factor for Q point*. That is, in a practical amplifier, the remaining resistor values are selected for the desired gain, impedance, and stability of the circuit; then the value of R_A is selected (or adjusted) to give a desired operating point.

The values of C_1 and C_2 are dependent upon the *low-frequency limit* at which the amplifier is to operate. C_1 forms a high-pass (or low-cut) *RC* filter with R_B. (Refer to Sec. 5-1.) Capacitor C_2 forms another filter with the input resistance of the following stage (or the load).

For a given resistance value, a lower frequency requires a larger capacitor value. Of course, if the resistance can be made larger (with the same desired frequency), the capacitor value can be reduced. Since two-junction transistors are low-impedance, current-operated devices, the coupling capacitors in two-junction amplifiers are generally large, that is, in relation to those of FET (and vacuum tube) amplifiers.

5-5.2. Emitter Bypass for Transistor Amplifier Stages

Figure 5-11 shows (in phantom) a bypass capacitor C_3 across emitter resistor R_E. This arrangement permits R_E to be removed from the circuit as far as the signal is concerned, but leaves R_E in the circuit (in regards to direct current). With R_E removed from the signal path, the voltage gain is approximately R_L divided by the dynamic resistance of the transistor, and the current gain is approximately equal to a-c β of the transistor. Thus, the use of an emitter-bypass capacitor permits the highly temperature stable d-c circuit to remain intact while providing a high signal gain.

An emitter-bypass capacitor also creates some problems. Transistor input impedance changes with frequency and from transistor to transistor, as does β. Thus, current and voltage gains can only be approximated. When the emitter resistance is bypassed, the circuit input impedance is approximately β times transistor input impedance, making circuit input impedance subject to variation and unpredictable.

The emitter bypass is generally used where maximum gain must be obtained from a single stage of amplification, and a stable gain is of little concern. The value of the emitter-bypass capacitor should be such that the *reactance is less than the transistor input impedance* at the lowest frequency of operation. This will effectively short the emitter (signal path) around R_E.

The value of C_3 can be found by

$$\text{Capacitance} = \frac{1}{6.2FR}$$

where capacitance is in farads; F is the low-frequency limit in hertz; and R is the maximum input impedance of the transistor in ohms.

5-5.3. Basic Audio Amplifier with Partially Bypassed Emitter

Figure 5-12 shows a basic single-stage two-junction transistor used as an audio amplifier with a partially bypassed emitter resistor. This circuit is a compromise between the basic amplifier without bypass and the fully bypassed emitter. The d-c characteristics of both the unbypassed and partially bypassed circuits are essentially the same. The circuit values (except C_3 and R_C) are calculated in the same way for both circuits. However, the voltage and current gains for a partially bypassed amplifier are greater than for an unbypassed circuit, but less than for the fully bypassed circuit.

The value of R_C is chosen on the basis of voltage gain, even though current gain will be increased when voltage gain increases. R_C should be substantially smaller than R_E. Otherwise, there will be no advantage to the partially bypassed circuit. However, a smaller value for RC will require a larger value for C_3, since the C_3 value is dependent upon the R_C value and the desired low-frequency cutoff point.

With the circuit of Fig. 5-12, current gain and voltage gain are dependent upon the ratios of R_B/R_C and R_L/R_C, respectively. Thus, the value of R_E has little or no effect on circuit gain.

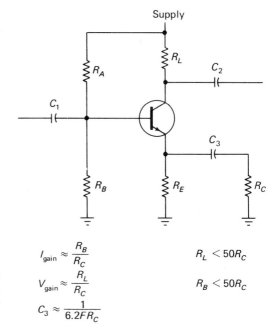

Figure 5-12 Basic single-stage two-junction transistor used as an audio amplifier with partially bypassed emitter resistor

$$I_{gain} \approx \frac{R_B}{R_C}$$

$$V_{gain} \approx \frac{R_L}{R_C}$$

$$C_3 \approx \frac{1}{6.2 F R_C}$$

$$R_L < 50 R_C$$

$$R_B < 50 R_C$$

5-6. BASIC FET AMPLIFIER STAGE

Figure 5-13 shows a basic single-stage FET amplifier. Input and output coupling capacitors C_1 and C_2 are added to prevent d-c flow to and from external circuits. Bypass capacitor C_3, connected across R_S, is required only under certain conditions. Also note that the circuit of Fig. 5-13 is the FET version of the Fig. 5-11 circuit. Thus, many of the circuit characteristics are similar.

One major difference is that voltage gain in the FET circuit (Fig. 5-13) is set by factors of R_L, R_S and y_{fs}. Another difference is that the values of coupling capacitors for FET circuits are much smaller than for two-junction transistors. This is because FETs are high-impedance, voltage-operated devices.

Sufficient feedback. The design of an FET amplifier stage can be checked by noting if there is sufficient feedback. Such a condition occurs when the calculated gain is at least 75 per cent of the R_L/R_S ratio; if so, there is sufficient feedback to be of practical value. As an example, assume that R_L is 30 kΩ, R_S is 3 kΩ, and y_{fs} is 7000 μmhos. The ratio of R_L/R_S is 10, with the gain slightly over 9.

$$\frac{R_L}{R_S} = \frac{30 \text{ k}\Omega}{3 \text{ k}\Omega} = 10; \quad \text{gain} = \frac{30 \text{ k}\Omega}{1/7000 \text{ μmho} + 3 \text{ k}\Omega} = 9+$$

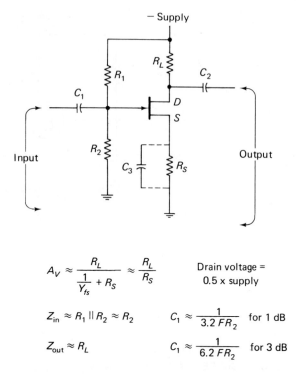

Figure 5-13   Basic common-source FET amplifier stage

$$A_v \approx \frac{R_L}{\frac{1}{Y_{fs}} + R_S} \approx \frac{R_L}{R_S} \qquad \text{Drain voltage} = 0.5 \times \text{supply}$$

$$Z_{in} \approx R_1 \| R_2 \approx R_2 \qquad C_1 \approx \frac{1}{3.2\, F R_2} \text{ for 1 dB}$$

$$Z_{out} \approx R_L \qquad C_1 \approx \frac{1}{6.2\, F R_2} \text{ for 3 dB}$$

Since 75 per cent of 10 is 7.5, the gain of 9 is greater, and there is sufficient feedback. Under these conditions the design should be stable.

5-6.1. Source Bypass for FET Amplifier Stage

Figure 5-13 shows (in phantom) a bypass capacitor C_3 across source resistor R_S. This arrangement permits R_S to be removed from the circuit as far as signal is concerned, but leaves R_S in the circuit (in regards to direct current). With R_S removed from the signal path, the voltage gain is approximately equal to $y_{fs} \times R_L$. Thus, the use of a bypass capacitor permits a temperature-stable d-c circuit to remain intact, while providing a high signal gain.

5-6.2. Other Single-Stage FET Amplifiers

Figures 5-14 through 5-18 show the circuit diagrams and design equations for some typical FET amplifier configurations. The following notes apply to these circuits.

Sec. 5-6   Basic FET Amplifier Stage

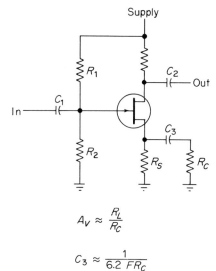

Figure 5-14   Basic FET audio amplifier with partially bypassed source resistor

$$A_V \approx \frac{R_L}{R_C}$$

$$C_3 \approx \frac{1}{6.2 \, FR_C}$$

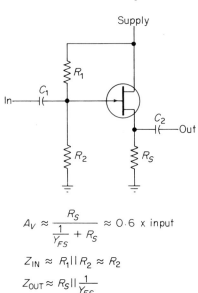

$$A_V \approx \frac{R_S}{\frac{1}{Y_{FS}} + R_S} \approx 0.6 \times \text{input}$$

$$Z_{IN} \approx R_1 \| R_2 \approx R_2$$

$$Z_{OUT} \approx R_S \| \frac{1}{Y_{FS}}$$

Figure 5-15   Basic common-drain (source-follower) FET audio-amplifier stage

The value of R_C in Fig. 5-14 is chosen on the basis of desired voltage gain. R_C should be substantially smaller than R_S. Otherwise, there will be no advantage to the partially bypassed design. As shown by the equations, voltage gain is approximately equal to R_L/R_C. This holds true unless both y_{fs} and R_C are very low (where $1/y_{fs}$ is about equal to R_C). In that case, a more accurate gain approximation is $R_L/(1/y_{fs}) + R_C$.

130 AF Amplifier Design Examples Chap. 5

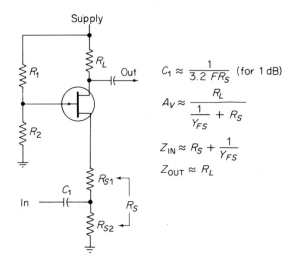

Figure 5-16 Basic common-gate FET amplifier stage

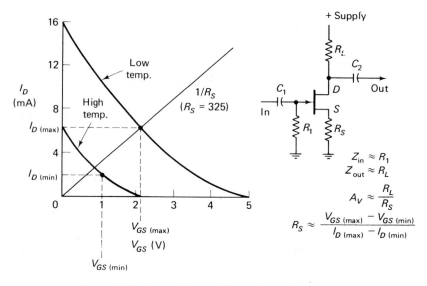

Figure 5-17 Basic FET amplifier without fixed bias

The circuit of Fig. 5-15 is used primarily where high-input impedance and low-output impedance (with no phase inversion) are required, but no gain is needed. The source follower is the FET equivalent of the two-junction transistor emitter follower and the vacuum tube cathode follower.

The circuit of Fig. 5-16 is used primarily where low-input impedance and high-output impedance (with no phase inversion) are required. Gain is

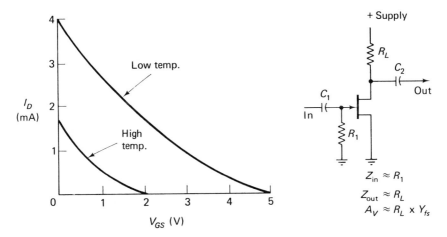

Figure 5-18 Basic FET amplifier with zero bias

determined (primarily) by circuit values, rather than FET characteristics. The common-gate amplifier is the FET equivalent of the two-junction common-base amplifier, and the vacuum tube common-grid amplifier.

The circuit of Fig. 5-17 has no fixed bias. The major difference in this circuit, and the FET amplifiers with fixed bias is that the amount of I_D at the Q point is set entirely by the value of R_S. It may not be possible to achieve a desired I_D with a practical value of R_S. Thus, it may not be possible to operate at the zero-temperature-coefficient point. If this is of less importance than minimizing the number of circuit components (elimination of one resistor), the circuit of Fig. 5-17 can be used in place of the fixed-bias FET amplifier.

The circuit of Fig. 5-18 is also without fixed bias and stabilizing feedback. Any FET will have some value of I_D at zero V_{GS}. If the FET has characteristics similar to those of Fig. 5-18, the I_D will vary between about 1.75 and 4 mA, depending upon temperature, and from FET to FET. Therefore, with a zero-bias circuit it is impossible to set the Q-point I_D at any particular value. Likewise, drain voltage Q point is subject to considerable variation. Since there is no source resistor, there is no negative feedback. Thus, there is no means to control this variation in I_D. For these reasons, the zero-bias circuit is used where circuit stability is of no particular concern.

5-7. MULTISTAGE TRANSISTOR AMPLIFIERS

When stable voltage gains greater than about 20 are required, and is not practical to bypass the emitter (or source) resistor of a single stage, two or more transistor amplifier stages can be used in *cascade* (where the output of one transistor is fed to the input of a second transistor). In theory, any number of two-junction or FET amplifiers can be connected in cascade to increase voltage gain. In practice, the number of stages is usually limited to three. The overall gain of the amplifier is equal (approximately) to the cumulative gain of each stage, multiplied by the gain of the adjacent stage.

As an example, if each stage of a three-stage amplifier has a gain of 10, the overall gain is (approximately) 1000 (10 × 10 × 10). Since it is possible to design a very stable single stage with a gain of 10, and adequately stable stages with a gains of 15 to 20, a three-stage amplifier could provide gains in the 1000 to 8000 range. Generally, this is more than enough voltage gain for most practical applications. Using the 8000 figure, a 1-μV input signal (say from a low-voltage transducer or delicate electronic device) can be raised to the 8-mV range, while maintaining stability in the presence of temperature and power supply variations.

5-7.1. Basic Considerations for Multistage Amplifiers

Any of the single-stage amplifiers described in previous sections of this chapter can be connected to form a two-stage or three-stage *voltage* amplifier. For example, the basic stage (without emitter or source bypass) can be connected to two like stages in cascade. The result is a highly temperature-stable voltage amplifier. Since each stage has its own feedback, the gain is precisely controlled and very stable.

It is also possible to mix stages to achieve some given design goal. For example, a three-stage amplifier can be designed using a highly stable, unbypassed amplifier for the first stage, and two bypassed amplifiers for the remaining stages. Assuming a gain of 10 for the unbypassed stage, and gains of 30 for the bypassed stages, this results in an overall gain of 9000. Of course, with the bypassed stages the gain is dependent upon the transistor characteristics (dynamic impedance, h_{fe}, y_{fs}, etc.), and is therefore unpredictable. However, once the gain is established for a given amplifier, the gain should remain fairly stable.

Since design of a multistage, capacitor-coupled voltage amplifier is essentially the same as for individual stages, no specific circuit example is given. In practical terms, each stage is analyzed and designed as described in previous sections of this chapter. However, a few precautions must be considered.

Distortion and clipping. As in the case with any high-gain amplifier, the possibility of *overdriving* a multistage transistor amplifier is always present. If the maximum input signal is known, check this value against the overall gain and the maximum allowable output signal swing.

As an example, assume an overall gain of 1000 and a supply voltage of 20 V. Typically, this implies a 10-V Q point (for the output collector or drain) and a 20-V (peak-to-peak) output swing (from 0 to 20 V). In practice, a swing from about 0 to 19 V is more realistic. Either way, a 20-mV (peak-to-peak) input signal, multiplied by a gain of 1000, will drive the final output to its limits, and possibly into distortion or clipping.

Feedback. When each stage of a multistage amplifier has its own feedback (local or stage feedback), the most precise control of gain is obtained. However, such feedback is often unnecessary. Instead, overall feedback (or loop feedback) can be used, where a part of the output from one stage is fed back to the input of a previous stage. Usually, such feedback is through a resistance (to set the amount of the feedback), and the feedback is from the final stage to the first stage. However, it is possible to use feedback from one stage to the next (second stage to first stage, third stage to second stage, etc.).

Feedback phase inversion. There is a problem of phase inversion when using loop or overall feedback. In a common-emitter or common-source amplifier, the phase is inverted from input to output. If feedback is between two stages, the phase is inverted twice, resulting in *positive feedback*. This usually produces oscillation. In any event, positive feedback will not stabilize gain. The phase inversion problem can be overcome, when stages are involved, by connecting the output collector (or drain) or the second stage back to the emitter (or source) of the first stage. This will produce the desired *negative feedback*.

As an example, if the base (or gate) of the first stage is swinging positive, the collector (or drain) of that stage will swing negative, as will the base (or gate) of the second stage. The collector (or drain) of the second stage will swing positive, and this positive swing can be fed back to the emitter (or source) of the first stage. A positive input at the emitter (or source) has the same effect as a negative at the base (or gate). Thus, negative feedback is obtained.

Low-frequency cutoff. Unless direct coupling is used, as described in Sec. 5-8, coupling capacitors must be used between stages, as well as at the input and output. Such capacitors form a low-pass *RC* filter with the base-to-ground (or gate-to-ground) resistance. Thus, each stage has its own low-pass filter. In multistage amplifiers, the *effects of these filters are cumulative*.

As an example, if each filter causes a 1-dB drop at some given cutoff frequency, and there are three filters (one at the input and two between stages), the result is a 3-dB drop at that frequency in the final output. If this cannot be tolerated, the *RC* relationship must be redesigned. In practical terms, this means increasing the value of *C*, since a change in *R* will usually produce some undesired shift in operating point or other circuit characteristic (such as input-output impedance).

5-8. DIRECT-COUPLED TRANSISTOR AMPLIFIERS

The amplifiers discussed thus far cannot amplify *direct currents* (or direct voltages) since such currents will not be passed by coupling capacitors or transformers. Also, amplifiers using transformers and/or coupling capacitors are not well suited for amplification of very low frequencies.

For reasons discussed in Sec. 5-2, direct-coupled amplifiers are used if direct currents and/or very low-frequency signals must be amplified. Direct-coupled amplifiers, also known as DC amplifiers, permit a signal to be fed directly to the transistor without the use of any coupling device. Direct-coupled amplifiers may be single stage or multistage. However, direct coupling is generally limited to three stages.

5-8.1. Practical Direct-Coupled Two-Junction Transistor Amplifier

Figure 5-19 is the working schematic of a two-stage direct-coupled *complementary amplifier*. That is, the output of an NPN is fed into a PNP transistor to increase stabilization. The increased stabilization for the

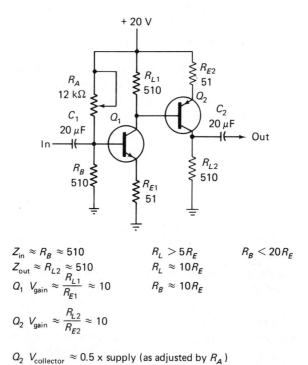

$Z_{in} \approx R_B \approx 510$ $R_L > 5R_E$ $R_B < 20R_E$
$Z_{out} \approx R_{L2} \approx 510$ $R_L \approx 10R_E$
$Q_1\ V_{gain} \approx \dfrac{R_{L1}}{R_{E1}} \approx 10$ $R_B \approx 10R_E$

$Q_2\ V_{gain} \approx \dfrac{R_{L2}}{R_{E2}} \approx 10$

$Q_2\ V_{collector} \approx 0.5 \times$ supply (as adjusted by R_A)

Figure 5-19 Basic direct-coupled complementary amplifier circuit

complementary direct-coupled amplifier arises from the fact that a change in collector current of Q_1 (due to temperature, power supply variation, etc.) is *opposed by an equal change* in collector current of Q_2 (but in the opposite direction). If more stages are to be added, the complementary system is continued. That is, NPN and PNP amplifiers are used alternately.

Two- and three-stage amplifiers, similar to that shown in Fig. 5-19, are available commercially as *hybrid circuits*. Such circuits are complete (or nearly complete) packages similar to integrated circuits. It is generally easier to design with hybrid circuits rather than with individual components, since impedance relationships, Q point, and so forth, have been calculated by the circuit manufacturer. Also, the datasheets supplied with the hybrid circuits provide information regarding source voltage, gain, impedances, and the like.

The datasheet information can be followed to adapt the hybrid circuit for a specific application. However, in some cases, it is necessary to select values of components external to the hybrid circuit package. For this reason, and since it may be necessary to design a multistage direct-coupled amplifier (with individual components) for some special application, the following design considerations and examples are provided.

Design considerations. The circuit elements for transistor Q_1 in Fig. 5-19 are essentially the same as for the circuit of Fig. 5-11 (the basic single-stage two-junction transistor amplifier). Also note that the same circuit arrangement is used for transistor Q_2, except that R_A and R_B are omitted (as is the coupling capacitor between stages).

The design considerations for the circuit of Fig. 5-19 are essentially the same as for the circuit of Fig. 5-11, except for the following:

The range of input voltage which may be applied to a direct-coupled transistor is small. The forward bias must not be increased to the point where the transistor operates in its saturated region, nor can the bias be reduced so that the transistor is cut off.

The input impedance of the complete circuit is approximately equal to R_B. The output impedance is set by R_{L2}.

The value of C_1 is dependent upon the low-frequency limit and the value of R_B. For example, if the low-frequency limit is 30 Hz, an approximate value of 20 μF is required for C_1 to produce 1 dB (at 30 Hz, with R_B at 510 Ω). The value of C_2 is dependent upon the low-frequency limit and the value of input resistance of the following stage (or the load). Capacitors C_1 and C_2 can generally be eliminated, except in those cases where an a-c signal is mixed with direct current.

When two stages are direct coupled in the stabilized circuit of Fig. 5-19, the overall voltage gain is about 70 per cent (possibly higher) of the combined gains of each stage. The gain of each stage is approximately 10, as set by the 10 to 1 ratio of collector emitter resistances. The combined gain of the circuit is theoretically 100, and practically about 70. Thus, an input of 100 mV is increased to about 7 V at the output.

The signal at the base of Q_2 is approximately 10 times the signal at the base of Q_1. Thus, there is an approximate 1-V signal at the base of Q_2. This signal can be a varying direct current or an a-c sinewave, or even a pulse. In any event, the base of Q_2 must be biased to accommodate the 1-V signal.

The collector of Q_2 should be approximately 10 V at the Q point. This will allow for the full 7-V output swing. With the collector of Q_2 at 10 V, there is an approximate 20-mA current through R_{L2} and R_{E2}. This produces an approximate 1-V drop across R_{E2}. Assuming that the transistors are silicon, the base of Q_2 should be about 0.5 V from the emitter. Since Q_2 is PNP, the base should be more negative (or less positive) than the emitter. The emitter of Q_2 is at about $+19$ V ($+20$-V supply, minus the 1-V drop across R_{E2}). Thus, the base of Q_2 should be $+18.5$ V. This also sets the collector voltage for Q_1 at the Q point. A Q-point collector voltage of 18.5 V will allow for the full 1-V signal swing.

The Q-point voltage of Q_1 is set by the collector current (approximately 3 mA). In turn, the collector current is set by the bias network R_A and R_B in the usual manner. In practice, the resistance values are approximated, and then R_A is adjusted to give the desired Q point. The final adjustment of R_A is made for distortion-free 7-V signal at the collector of Q_2 (with a 100-mV input signal applied to Q_1).

Since gain of the amplifier is set by circuit values, little concern need be given to transistor β. Of course, the β of each transistor must be greater than ten at all frequencies that the circuit must amplify. Note that the circuit of Fig. 5-19 is highly stable, and has a very wide band frequency response (typically from direct current on up to whatever maximum is set by the transistor high-frequency limitations).

5-8.2. Darlington Compounds

Figure 5-20 shows the basic Darlington circuit (known as the Darlington compound), together with two other versions of the circuit. As shown, the Darlington compound is an emitter follower (or common collector) driving a second emitter follower. Going back to the basic amplifier theory, an emitter follower provides no voltage gain, but can provide considerable power gain.

The main reason for using a Darlington compound (especially in audio work) is to produce high-current (and power) gain. For example, Darlington compounds are often used as *audio drivers* to raise the power of a signal from a voltage amplifier to a level suitable to drive a final power amplifier. Darlingtons are also used as a substitute for a driver section (or to eliminate the need for a separate driver).

Sec. 5-8 Direct-Coupled Transistor Amplifiers 137

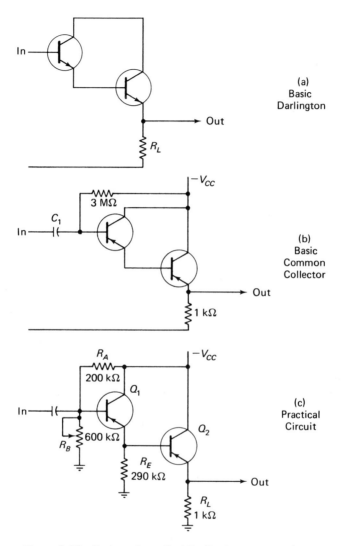

Figure 5-20 Basic and practical Darlington compounds

Emitter-follower Darlington. When the Darlington is used as a common collector, as shown in Fig. 5-20, the output impedance is approximately equal to the load resistance R_L. The input impedance is approximately equal to $\beta^2 \times R_L$. The current gain is approximately equal to the average β of the two transistors, squared. However, in most common collectors (emitter followers), power gain is of primary concern. That is, the designer is interested in how much the signal power can be increased across a given output load.

As an example, assume that the value of R_L (in Fig. 5-20(c)) is 1 kΩ and that the average β is on the order of 15. This results in an input impedance of about 225 kΩ ($15^2 \times 1000$), and an output impedance of 1 kΩ. Now assume that a 2.5-V signal is applied at the input, and an output of 2 V appears across R_L. This input power is $2.5^2/225$ kΩ ≈ 0.028 mW. The output power is $2^2/1$ kΩ ≈ 4 mW. The power gain is $4/0.028 \approx 140$.

Common-emitter Darlington. Darlington compounds can be used as common-emitter amplifiers to provide voltage gain. This can be accomplished by simply adding a collector resistor to any of the circuits in Fig. 5-20, and taking the output from the collector rather than the emitter. In effect, Q_1 then becomes a common collector driving Q_2, which appears as a common-emitter amplifier. The entire circuit then appears as a common-emitter amplifier and can be used to replace a single transistor. Such an arrangement is often used when *high voltage gain* is desired.

A more practical method of using a Darlington as a common-emitter amplifier is to eliminate R_B and R_E (of Fig. 5-20(c)), ground the emitter of Q_2, and transfer R_L to the collector of Q_2. Such an arrangement is shown in Fig. 5-21. The circuit of Fig. 5-21 is stabilized by the collector feedback through R_A, which holds both collectors at a potential somewhat less than 0.5 V from the base of Q_1. Note that both collectors are at the same voltage, and that this voltage is approximately equal to two base-emitter voltage drops (or about 1.5 V for two silicon transistors).

With the circuit of Fig. 5-21, the current gain is approximately equal to the ratio of R_A/R_L. Both the input and output voltage swings are somewhat limited in the Fig. 5-21 circuit. The input is biased at approximately 1 to 1.5 V. However, voltage gains of 100 (or more) are possible, since input

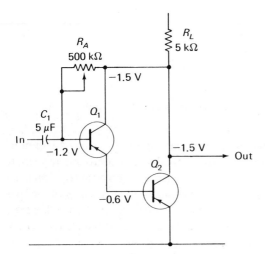

Figure 5-21 Darlington compound with collector feedback and common-emitter output

impedance (or resistance) is approximately equal to the ratio of R_A/current gain, or equal to R_L. (With input and output impedances approximately equal to R_L, the voltage gain follows the current gain.)

Multistage Darlingtons. Darlington compounds need not be limited to two transistors. Three (and even four) transistors can be used in the Darlington circuit. A classic example of this is the General Electric circuit of Fig. 5-22 (which is available in hybrid form, in a TO-5 style package). This circuit is essentially a common-collector and common-emitter Darlington, followed by a common-emitter amplifier.

With R_I out of the circuit, both the input and output impedances are set by R_L. (In practice, the input impedance is slightly higher than R_L, typically on the order of 700 to 800 Ω). With R_I removed, the voltage gain is about 1000. When R_I is used, the input impedance is approximately equal to R_I, and the voltage gain is reduced accordingly. For example, if R_I is 10 kΩ, the 1000 voltage gain drops to about 50.

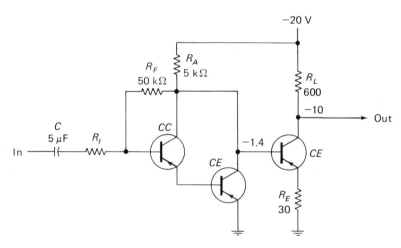

Figure 5-22 General Electric *CC-CE-CE* multistage Darlington compound

5-8.3. Practical Direct-Coupled FET Amplifier

Both MOSFETs and JFETs can be used in direct-coupled amplifiers. However, MOSFETs are especially well suited to direct-coupled applications. Since the gate of a MOSFET acts essentially as a capacitor, rather than a diode junction, no coupling capacitor is needed between stages. For a-c signals, this means that there are no low-frequency cutoff problems, *in theory*. In practical design, the input capacitance can form an *RC* filter with the source resistance, and result in some low-frequency attenuation.

140 AF Amplifier Design Examples Chap. 5

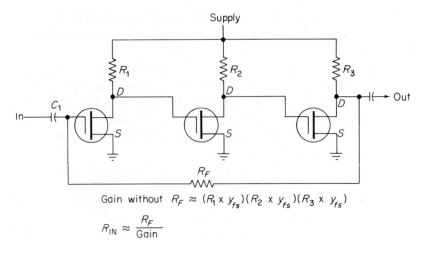

Gain without $R_F \approx (R_1 \times y_{fs})(R_2 \times y_{fs})(R_3 \times y_{fs})$

$R_{IN} \approx \dfrac{R_F}{\text{Gain}}$

Figure 5-23 All MOSFET three-stage amplifier

Figure 5-23 is the working schematic of an all-MOSFET three-stage amplifier. Note that all three MOSFETs are of the same type, and all three drain resistors (R_1, R_2 and R_3) are the same value. This arrangement simplifies design. At first glance it may appear that all three stages are operating at zero bias. However, when I_D flows, there is some drop across the corresponding drain resistor, producing a voltage at the drain of the stage, and an identical voltage at the gate of the next stage. The gate of the first stage is at essentially the same voltage as the drain of the last stage, because of feedback resistor R_F. There is no current drain through R_F, with the possible exception of reverse gate current (which can be ignored for practical purposes).

Operating point. To find a suitable operating point for the amplifier, it is necessary to trade off between desired output voltage swing, MOSFET characteristics, and supply voltage. For example, assume that an output swing of 7-V peak-to-peak is desired, the supply voltage is 24 V, and that the I_D is about 0.55 mA when V_{GS} is 7 V. A suitable operating point would be 7 V to accommodate the 7-V output swing without distortion. The swing would be from 3.5 to 10.5, about the 7-V point.) This requires an 18-V drop from the 24-V supply. With 0.55-mA I_D and an 18-V drop, the values of R_1, R_2 and R_3 should be about 33 kΩ.

Gain. The overall voltage gain is dependent upon the relationship of the gain without feedback and the feedback resistance R_F. Gain without feedback is determined by y_{fs} and the value of R_1, R_2, and R_3. For example, assume a

y_{fs} of 1000 μmhos (0.001 mho); the gain of each stage is 33 (33 × 0.001 = 33). With each stage at a gain of 33, the overall gain (without feedback) is about 36 kΩ.

To find the value of R_F, divide the gain without feedback by the desired gain. Multiply the product by 100; then multiply the resultant product by the value of R_1. For example, assume a desired gain of 3000 (the gain without feedback is 36,000): 36,000/3000 = 12; 12 × 100 = 1200, 1200 × 33,000 = 39.6 MΩ (use the nearest standard to 40 MΩ).

Input impedance. The input impedance is dependent upon the relationship of gain and feedback resistance. The approximate impedance is R_F/gain.

Since gain is dependent upon y_{fs}, input impedance is subject to variation from FET to FET, and with temperature.

Direct-current amplification. The circuit of Fig. 5-23 requires one coupling capacitor at the input. This is necessary to isolate the input gate from any direct-current voltage that may appear at the input. This makes the circuit of Fig. 5-23 unsuitable for use as a d-c amplifier. The circuit can be converted to a d-c amplifier when the coupling capacitor is replaced by a series resistor R_{IN}, as shown in Fig. 5-24.

The considerations concerning operating point are the same for both circuits. However, the series resistance must be terminated at a d-c level equivalent to the operating point. For example, if the operating point is −7 V, point A must be at −7 V. If point A is at some other d-c level, the operating point is shifted.

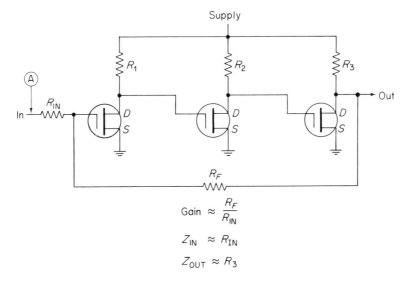

Figure 5-24 All MOSFET three-stage direct-coupled amplifier

The relationships of input impedance, R_F, and gain still hold. However, input impedance is approximately equal to R_{IN}. Therefore, gain is approximately equal to the ratio of R_F/R_{IN}. This makes it possible to control gain by setting the R_F/R_{IN} ratio. Of course, the gain cannot exceed the gain-without-feedback (open-loop) factor, no matter what the ratio of R_F/R_{IN} (closed-loop). As a general rule, the greater the ratio of open-loop gain to closed-loop gain, the greater the circuit stability.

As an example, assume that the open-loop gain is 36,000, and the desired gain is 6000. This requires a ratio of 6 to 1.

As another example, assume that the desired gain is 5000, R_F is 40 MΩ, and the open-loop gain is 36,000. 5000 is considerably less than 36,000, so the circuit is well capable of producing the desired gain with feedback. To find the value of R_{IN}

$$R_{IN} \approx \frac{40 \times 10^6}{5 \times 10^3} \approx 8 \times 10^3 \; (8 \text{ k}\Omega)$$

Amplification from grounded sources. The circuit of Fig. 5-24 requires that the signal source be at a d-c level equal to the operating point. In many cases, it is necessary to amplify d-c signals at the zero or ground level. This can be accomplished by using a depletion-mode JFET at the input, as shown in Fig. 5-25.

Feedback is introduced by connecting the sources of both Q_1 and Q_3 to a common-source resistor R_2. The source of Q_2 is not provided with a source resistor, but there is some bias on Q_2 produced by the I_D drop across R_3. The input impedance of the Fig. 5-25 circuit is set by the value of R_1 and the gate-drain capacitance of Q_1.

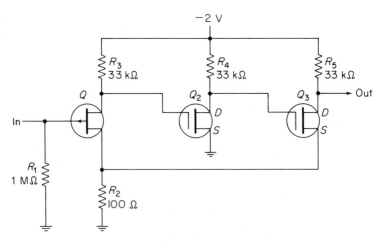

Figure 5-25 Direct-coupled amplifier with JFET input, followed by two MOSFET stages

Gain can be sacrificed for stability by increasing the value of R_2. With the values shown and typical FETs, the gain should be on the order of 3000 to 5000. This bias and operating point for Q_2 and Q_3 is set by R_3, shown as 33 kΩ. In practice, the value of R_3 is approximated by calculation, and then adjusted for a desired operating point at the output (drain) of Q_3.

5-8.4. Combination FET and Two-Junction Amplifiers

In certain applications, FET stages and two-junction transistor stages can be combined. The classic example is where a single FET stage is used at the input, followed by two two-junction amplifier stages. Such an arrangement takes advantage of both the FET and two-junction transistor characteristics.

An FET is a voltage-operated device permitting large voltage swings with low currents. This makes it possible to use high-resistance values (resulting in high impedances) at the input and between stages. In turn, these high-resistance values permit the use of low-value coupling capacitors and eliminate the need for bulky, expensive electrolytic capacitors. If operated at the 0-TC point, the FET is highly temperature stable, tending to make the overall amplifier equally stable. However, FETs have the characteristic of operating at low currents, and are therefore considered as low-power devices.

Two-junction transistors are essentially current-operated devices, permitting large currents at about the same voltage levels as the FET. Thus, with equal supply voltage and signal current swings, the two-junction transistor can supply much more current gain (and power gain) than the FET. Since currents are high, the impedances (input, interstage, and output) must be low in two-junction transistor amplifiers. This requires large-value coupling capacitors if low frequencies are involved. The low impedances also place a considerable load on devices feeding the amplifier, particularly if the devices are high impedance. On the other hand, a low-output impedance is often a desirable characteristic for an amplifier.

When a FET is used as the input stage, the amplifier input impedance is high. This places a small load on the signal source, and allows the use of a low-value input coupling capacitor (if required). If the FET is operated at the 0-TC point, the amplifier input is temperature stable. (Generally, the input stage is the most critical in regards to temperature stability.) When two-junction transistors are used as the output stages, the output impedance is low, and current gain (as well as power gain) is high.

Combination amplifiers can be direct-coupled or capacitor-coupled, depending on requirements. The direct-coupled configuration offers the best low-frequency response, permits d-c amplification, and is generally simpler (uses less components). The capacitor-coupled combination amplifier permits a more stable design, and eliminates the voltage regulation problem (that is, a d-c amplifier cannot distinguish between signal level changes and power supply changes).

The FET can be combined with any of the classic two-stage two-junction transistor amplifier combinations. The two most common combinations are the Darlington pair (for no voltage gain, but high-current gain and low-output impedance), and the NPN-PNP complementary pair amplifier (for both current gain and voltage gain).

FET input, two-junction transistor output amplifier. Figure 5-26 is the working schematic of a direct-coupled amplifier using an FET input stage and a two-junction transistor pair as the output. Note that local feedback is used in the FET stage (provided by R_S), as well as overall feedback (provided by R_4).

Input impedance is set by the value of R_2, as usual. Output impedance is set by the combination of R_4 and R_S. Since R_S is quite small in comparison to R_4, the output impedance is essentially equal to R_4.

The gain of the FET stage is set by the ratio of R_L to R_S, plus the $1/y_{fs}$ factor. Since R_S is quite small, the FET gain is set primarily by the ratio of R_L to $1/y_{fs}$. The gain of the two-junction transistor pair is set by the β of the two transistors and the feedback. Thus, the gain can only be estimated.

The drop across R_3 is the normal base-emitter drop of a transistor (about 0.5 to 0.7 V for silicon, and 0.2 to 0.3 V for germanium). The drop across R_L

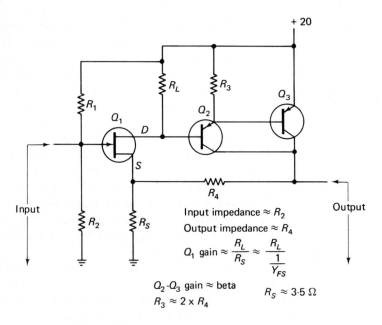

Figure 5-26 Direct-coupled hybrid amplifier

is twice this value (about 1 to 1.5 V for silicon, and 0.4 to 0.6 V for germanium).

For a typical silicon transistor, the base of Q_2 and the drain of Q_1 operate at about 1 V removed from the supply. In a practical experimental circuit, R_L must be adjusted to give the correct bias for Q_2 (and operating point for Q_1). The same is true for R_3. As a first trial value, R_3 should be about twice the value of R_4.

Design starts with selection of I_D for the FET. If maximum temperature stability is desired, use the 0-TC level of I_D. This usually requires a fixed bias. If temperature stability is not critical, the FET can be operated at zero bias by omitting R_1. There is some voltage developed across R_S. However, since R_S is small, the V_{GS} is essentially zero, and the I_D is set by the 0-V_{GS} characteristics of the FET.

With the value of I_D set, select a value of R_L that produces approximately a 1- to 1.5-V drop to bias Q_2.

The input impedance is set by R_2, with the output impedance set by R_4. The value of R_3 is approximately twice that of R_4. The value of R_S is less than 10 Ω, typically on the order of 3 to 5 Ω.

As a brief design example, assume that the circuit of Fig. 5-26 is to provide an input impedance of 1 MΩ, an output impedance of 500 Ω, and maximum gain. Temperature stability is not critical.

Under these conditions, the values of R_2 and R_4 are set at 1 MΩ and 500 Ω, respectively. For practical purposes, with the voltage drop across R_S ignored, Q_1 operates at $V_{GS} = 0$. Assume that I_D is 0.2 mA under these conditions. With the required drop of 1.5 V, and 0.2-mA I_D, the value of R_L is approximately 7.5 kΩ. Since R_4 is 500 Ω, R_3 should be 1 kΩ (as indicated by the equations).

The key component in setting up this circuit is R_L. With the circuit operating in experimental form, adjust R_L for the desired Q-point voltage at the output (collectors of Q_2 and Q_3).

Nonblocking direct-coupled amplifier. Generally, a direct-coupled amplifier does not require any coupling capacitors. One exception is a coupling capacitor at the input to isolate the amplifier from direct current (when the signal is composed of direct current and alternating current). When a coupling capacitor is used at the gate of a JFET (or at the base of some two-junction transistors) a condition known as *blocking* could occur.

Blocking is produced by the fact that the gate junction of a JFET is similar to that of a diode (as is the base junction of a two-junction transistor). The "diode" acts to rectify the incoming signal. If a capacitor is connected in series with the "diode", large signals can charge the capacitor. On one half-cycle, the diode is forward-biased and charges rapidly. On the opposite half-cycle, the diode is reverse-biased and discharges slowly. If the signal and charge are large enough, the amplifier can be biased at or beyond cutoff, until the capacitor discharges. Thus, the amplifier can be blocked to incoming signals for a period of time.

One method for eliminating the blocking condition is to use a MOSFET at the input. Since the gate of a MOSFET acts essentially as a capacitor (rather than a diode junction), there is no rectification of the signal and no blocking.

5-9. MULTISTAGE TWO-JUNCTION TRANSISTOR AMPLIFIER WITH TRANSFORMER COUPLING

Audio amplifier stages can be coupled by means of transformers. Iron-core transformers are used in the AF range, particularly when power amplification is required.

As with any coupling method, transformers impose certain problems and have certain advantages. These factors can be traded off to meet a specific design need.

Inductive reactance. One major problem with transformers is the inductive reactance created by the transformer windings. Inductive reactance increases with frequency; at frequencies beyond about 20 kHz, the inductive reactance of iron-core transformers becomes so high that signals cannot pass, or are greatly attenuated. For this reason, iron-core transformers are not used at higher frequencies. At low frequencies, the reactance drops to near zero, even with iron cores. Since transformers are placed across (or in shunt with) the amplifier circuits, the transformer winding acts as a short at low frequencies. If an amplifier must operate at very low frequencies (below about 20 Hz), transformers are not used.

Size and weight. For a transformer to handle any large amounts of power or current, the winding wire must be large. Also, the iron core must be large. Both of these add up to bulk weight. Except in certain cases, the added weight of transformers defeats the purpose of compact, lightweight, transistorized equipment.

Short-circuit burnout. Another problem with transformers is the danger of a short-circuit output, resulting in excessive, simultaneous voltage and current in the transistor collector. There is very little voltage drop across a transformer winding in a transistor collector circuit (compared with the drop across the load in an *RC* or direct-coupled amplifier.) If a short circuit in the output (say due to a short across the load) causes heavy current flow, the transistor can be damaged. The destructive condition is known as *secondary breakdown* or *second breakdown*. Secondary breakdown can be prevented (or minimized) by limiting the collector current-voltage product.

Another overload problem is presented when the load is a loudspeaker. High-fidelity loudspeaker systems can appear *capacitive* or *inductive*, as well as resistive. The current and voltage appearing in the amplifier will thus be out-of-phase when the load appears *reactive*. A 60° phase shift is not uncommon. At a 60° phase shift, half the supply voltage and all of the load current can appear simultaneously at the output transistor, or the full source voltage and one-half the load current can appear, depending on whether the load is capacitive or inductive.

As a guideline, there will be no secondary breakdown if the output transistor can dissipate *twice the power output* of the amplifier (for single-ended transformer amplifiers). If the amplifier is push pull, each output transistor must be capable of dissipating the full power output, if secondary breakdown is to be avoided.

Impedance matching. One of the major advantages to transformer coupling is the impedance-matching capability. The output impedance of a typical *RC* amplifier is on the order of several hundred (or thousand) ohms (generally set by the output collector resistor value). In audio systems (particularly for voice and music reproduction) this large output impedance must be matched to 4-, 8-, and 16-Ω loudspeaker systems. The severe mismatch results in power loss. With a transformer, the primary winding can be designed (or selected) to match the transistor circuit output impedance with the transformer secondary matching the loudspeaker (or other load) impedance.

Low supply voltage. Another major advantage of transformer coupling is caused by the low-voltage drop across the transformer winding. Because of this low-voltage drop, it is possible to operate a transformer-coupled amplifier with a much lower supply voltage than with an *RC* amplifier. As a guideline, the transformer-coupled amplifier can be operated at *one-half the supply voltage* required for a comparable *RC* amplifier.

Typical uses. Transformers are often used in the audio-amplifier sections of transistorized radio receivers and portable hifi systems. In these applications, very little power is required, so the transformers can be made compact and lightweight. Since there is no loss in impedance match and low voltage is required, the transformer-coupled circuits are ideal for battery operation. Transformers are also used in high-power, hifi-stereo systems and television audio sections where the added weight is of little consequence.

5-9.1. Circuit Analysis

Figure 5-27 is the working schematic for a classic transformer-coupled audio amplifier. The circuit has a class A input or *driver stage* and a class B push-pull output stage. The class A stage provides both voltage and

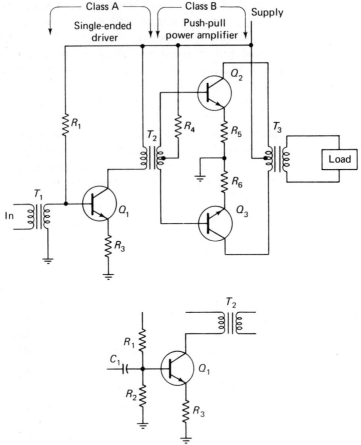

Alternate driver with *RC* input

Figure 5-27 Transformer-coupled audio amplifier

power amplification as needed to raise the low-input signal to a level suitable for the class B power output stage.

The class A stage can be transformer-coupled or *RC*-coupled at the input, as needed. Transformer coupling is used at the input where a specific impedance-match problem must be considered in design. The class A input stage can be driven directly by the signal source, or can be used with a preamplifier for very low-level signals. When required, a high-gain voltage amplifier (such as described in other sections of this chapter) is used as a *preamplifier*.

The push-pull output stage may be operated as a class B amplifier. That is, the transistors are cut off at the Q point and draw collector current only in the presence of an input signal. Class B is the most efficient operating mode for audio amplifiers, since it draws the least amount of current (and no current where there is no signal). However, true class B operation can result in *crossover distortion*.

The effects of crossover distortion can be seen by comparing the input and output waveforms on Fig. 5-28. In true class B operation, the *transistor remains cutoff* at very low-signal inputs (because transistors have low-current gain at cutoff) and turns on abruptly with a large signal. As shown on Fig. 5-28(a), there is no conduction when the base-emitter voltage V_{BE} is below about 0.65 V (for a silicon transistor). During the instantaneous pause when one transistor stops conducting and the other starts conducting, the output waveform is distorted.

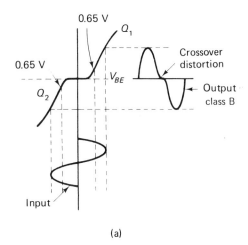

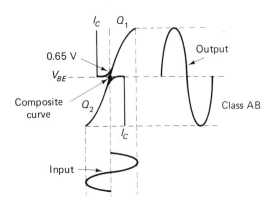

Figure 5-28 Effects of crossover distortion, and how it is eliminated by forward bias of the base-emitter junction

Distortion of the signal is not the only bad effect of this crossover condition. The instantaneous cutoff of collector current can set up large voltage transients equal to several times the size of the supply voltage. This can cause the transistor to break down.

Crossover distortion can be minimized by operating the output stage as class AB (or somewhere between B and AB). That is, the transistors are forward-biased just enough for a small amount of collector current to flow at the Q point. Some collector current is flowing at the lowest signal levels, and there is no abrupt change in current gain. The effects of this are shown in Fig. 5-28(b). The combined collector currents result in a *composite curve* that is essentially linear at the crossover point. This produces an output that is a faithful reproduction of the input, at least as far as the crossover point is concerned. Of course, class AB is less efficient than class B, since more current must be used.

Some designers use an alternative method to minimize crossover distortion. This technique involves putting diodes in series with the collector or emitter leads of the push-pull transistors. Because the voltage must reach a certain value (typically 0.65 V for silicon diodes) before the diode will conduct, the collector current curve is rounded (not sharp) at the crossover point.

Amplifier efficiency. The efficiency of an amplifier is determined by the ratio of collector input-to-output power. An amplifier with 70 per cent efficiency will produce 7-W output for a 10-W input (with power input being considered as collector source voltage multiplied by total collector current).

Typically, class B amplifiers can be considered as 70 to 80 per cent efficient. Class A amplifiers are typically in the 35 to 40 per cent efficiency range, with class AB amplifier showing 50 to 60 per cent efficiency.

In all cases, any amplifier circuit that produces an increase in collector current at the Q point will produce correspondingly lower efficiency. This results in a tradeoff between efficiency and distortion.

The efficiency produced by a class of operation also affects the heat-sink requirements. Any design that produces more collector current at the Q point requires a greater heat-sink capability. As a guideline, a *class A amplifier requires double the heat-sink capability of a class B*, all other factors being equal.

Distortion of class B versus class AB. In practical terms, push-pull amplifiers are usually designed for true class B operation (transistors at or near cutoff at the Q point), and then are tested (in experimental form) for distortion. If the distortion is severe, the base-emitter forward bias is increased so that some current flows at the Q point. Then the amplifier is tested again for distortion. The amount of forward bias is adjusted until the desired distortion is reached, or until there is a compromise between distortion and power output. This approach is usually more realistic than trying to design an amplifier for a given class of operation.

Output power. The true output power of an audio amplifier can be determined by measuring the voltage across the load, and then solving the equation

$$\text{Power output} = \frac{\text{voltage}^2}{\text{load impedance}}$$

The output of a transformer-coupled amplifier is also related to the collector current (produced by the signal) and the primary impedance of the transformer. In a push-pull amplifier, the relationship is

$$\text{Power output} = \frac{\text{current}^2 \times \text{primary impedance}}{8}$$

In the single-ended amplifier, the relationship is

$$\text{Power output} = \frac{\text{current}^2 \times \text{primary impedance}}{2}$$

These relationships provide a basis for design of transformer-coupled transistor amplifiers.

Transformer characteristics. Audio transformers are listed by primary impedance, secondary impedance, and power-output capability. From a practical standpoint, it is not always possible to find an off-the-shelf transformer with exact primary-secondary relationships at a given power rating. Most manufacturers will produce transformers with exact impedance relationships on a special-order basis. However, this is usually not practical, except for special applications. Instead, the transformer is generally selected for exact secondary impedance and the *nearest value* of primary impedance within the given power rating.

The following rules can be applied to selection of the transformers shown in Fig. 5-27.

Push-pull output transformer T_3. The rated power capability of transformer T_3 should be about 1.1 times the desired power. The secondary impedance should match the load impedance (loudspeaker or other) into which the amplifier must operate. The primary impedance should be determined by maximum collector current passing through the primary windings and the total collector voltage swing. The relationship is

$$\text{Primary impedance} = \frac{4(\text{supply voltage} - \text{min voltage})}{\text{max current}}$$

Minimum voltage can be determined by reference to the collector voltage-current curves for the transistors. As shown in Fig. 5-29, the minimum voltage point should be selected so that it is just to the right of the curved portion of the characteristics (where the curves start to straighten out). If curves are not available for the transistors, an arbitrary 2 V can be used for the minimum voltage. This is satisfactory for most power transistors.

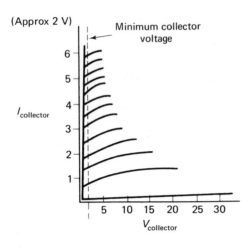

Figure 5-29 Locating minimum collector voltage point on typical power transistor curves

Maximum current can be determined by the total collector voltage swing and the desired power output. The relationship is

$$\text{Max current} = \frac{2.1 \times \text{power output}}{(\text{supply voltage} - \text{min voltage})}$$

As an example, assume that the supply voltage is 20 V, the desired output is 30 W, and the minimum voltage is an arbitrary 2 V.

$$\text{Max current} = \frac{2.1 \times 30}{(20 - 2)} = 3.5 \text{ A}$$

$$\text{Primary impedance} = \frac{4(20 - 2)}{3.5} \approx 21 \text{ }\Omega$$

Push-pull input transformer T_2. Transformer T_2 is the input transformer for the push-pull output stage and the output transformer for the single-ended driver. The secondary impedance should be chosen to match the signal input impedance of the push-pull stage, whereas the primary of T_2 is chosen to match the driver output.

The rated capability of T_2 should be equal to the input power of the driver stage (which is generally about three times the output to the push-pull stage).

The total secondary impedance of T_2 should be four times the signal impedance of Q_2 and Q_3. This input impedance is found by dividing the signal voltage by signal current.

The signal voltage and current are dependent upon the desired amount of collector signal current. In turn, the collector signal current is dependent upon the desired output power and the primary impedance of T_3 (total primary impedance). The relationship is

$$\text{Collector signal current} = \sqrt{\frac{8 \times \text{power output}}{\text{total primary impedance}}}$$

With the required collector signal current established, the input signal voltage and current for Q_2 and Q_3 can be found by reference to the *transfer characteristic curves*. Transfer curves are usually provided on power transistor datasheets. Typical transfer characteristics for power transistors are shown in Fig. 5-30. These characteristics illustrate the required base-emitter voltage (or signal voltage) and base current (or signal current) for a given collector current.

An additional factor must be considered in establishing signal voltage. The base-emitter voltages shown in Fig. 5-30 can be considered as the signal voltages only when the emitters of Q_2 and Q_3 are connected directly to ground. If emitter resistors (R_5 and R_6) are used to provide feedback stabilization, the voltage developed across the emitter resistors must be added to the base-emitter voltage to find a true signal voltage.

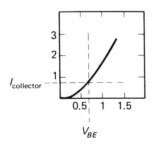
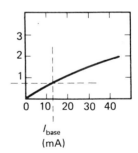

Figure 5-30 Curves showing base voltage and current versus collector current for typical power transistor

The values of R_5 and R_6 are chosen to provide a voltage drop *approximately equal* to the base-emitter voltage when the collector signal current is passing through the emitter resistors. Since the base-emitter voltage of a silicon power transistor is typically less than 1 V, the values for R_5 and R_6 can arbitrarily be set to provide a 0.5- to 1-V drop with normal collector signal current.

The primary impedance of T_2 *should match the output of* Q_1. The relationship is

$$\text{Primary impedance} = \frac{\text{supply voltage} - \text{min voltage}}{\text{collector current at } Q \text{ point}}$$

However, for practical design it is generally easier to select the transformer on the basis of secondary impedance and power rating, and then adjust the Q_1 output to match a primary impedance available with an off-the-shelf transformer.

A large primary impedance requires a high supply voltage for a given collector current. It may be necessary to trade off between supply voltage, available primary impedances (for off-the-shelf transformers with correct secondary impedance and power rating), and collector current.

Once the primary impedance of T_2 is established, the collector signal current for Q_1 can be determined. The relationship is

$$\text{Collector signal current} = \frac{2 \times \text{power output}}{\text{primary impedance}}$$

The power output from Q_1 must equal the required power input for Q_2 and Q_3. This power is determined by

$$\text{Power input} = \text{signal voltage} \times \text{signal current}$$

Input transformer T_1. As shown in Fig. 5-27, transformer T_1 can be omitted if the *RC* circuit is used. Generally, the transformer-coupled circuit is used if it is particularly important to match the impedance of the signal source to the amplifier. When used, the primary impedance of T_1 should equal that of the signal source. The secondary of T_1 should match the signal input of Q_1. The signal input impedance is found by

$$\text{Signal input impedance} = \frac{\text{signal voltage}}{\text{signal current}}$$

The signal (input) voltage and current are dependent upon the desired amount of collector signal current for Q_1 (previously established in calculating the secondary impedance of T_2). With the required collector signal current established, the input signal voltage and current for Q_1 can be found by reference to the transfer characteristic curves.

If emitter resistor R_3 is used to provide feedback stabilization, the voltage developed across the emitter resistor must be added to the base-emitter voltage to find a true signal voltage. The value of R_3 can be selected to provide between 0.5- and 1-V drop with normal collector signal current (as discussed for emitter resistors R_5 and R_6).

If transfer characteristics are not available for Q_1 (as is sometimes the case for low-power transistors), it is still possible to find the approximate signal voltage and current required to produce the necessary output. Signal voltage can be approximated by assuming that the base-emitter drop is 0.5 V. This must be added to any drop across emitter resistor R_3 (arbitrarily chosen to be between 0.5 and 1 V). Signal current can be approximated by dividing the desired collector signal current by β of Q_1. Of course, since β is variable, the input signal current can only be a rough estimate.

Supply voltage. The source voltage required for Q_1 is approximately double that required for Q_2 and Q_3. However, since all three transistors are operated from the same supply, the source voltage is chosen to match the requirements of Q_1. As a guideline,

3 to 9 V is used for power outputs up to 2 W

6 to 15 V for power up to 20 W

15 to 50 V for power up to 50 W

These rules apply to the output of Q_1. Transistors Q_2 and Q_3 require approximately one-half the voltage. For example, if Q_1 is to deliver 20 W (by itself), 6 to 15 V is required. If Q_2 and Q_3 are to deliver the same 20 W, the voltage required will run between 3 and 7.5 V.

In any amplifier circuit, a higher supply voltage permits lower currents (and vice versa), all other factors being equal.

Transistor selection. Both frequency limit and power dissipation must be considered in selecting transistors. In any audio system the transistors should have an f_{ae} higher than 20 kHz and preferably higher than 100 kHz. The power dissipation of Q_1 should be at lesst 3 times the required output of Q_1. The power dissipation of Q_2 and Q_3 should be between 1.3 and 1.5 times the required output of the amplifier circuit (at T_3). If any power dissipation exceeds about 1 W, heat sinks will be required.

Base bias resistances. Resistors R_1, R_2, and R_4 serve to drop the supply voltage to a level required at the transistor bases.

The drop across R_4 should (in theory) bias Q_2 and Q_3 at cutoff. That is, no base current should flow except in the presence of a signal. This is not true in a practical circuit, since the complete absence of base current will result in no drop across R_4. Also, such a bias condition will usually result in crossover distortion.

As a starting point for Q_2 and Q_3 bias values, assume that a small residual base current is going to flow in each transistor. The value of this residual current can be assumed (arbitrarily) as 0.1 times the normal signal current. The combined residual currents of Q_2 and Q_3 flow through R_4. The value of

R_4 is calculated by

$$R_4 = \frac{\text{supply voltage}}{2 \times \text{residual current}}$$

If transformer T_1 is used, resistor R_2 is omitted. Under these conditions, the drop across R_1 should bias Q_1 at the operating point. The base current of Q_1 flows through R_1 and produces the drop. The base current used to calculate the Q_1 input impedance can be used as a starting point for the R_1 value. The relationship is

$$R_1 = \frac{\text{supply voltage}}{\text{base current}}$$

If transformer T_1 is not used, resistor R_2 is added. The value of R_2 determines the *approximate* input impedance of the amplifier, and is generally so selected. Typically, R_2 is greater than 500 Ω, and less than 20 kΩ. When R_2 is used, the voltage drop from base to ground will also appear across R_2, resulting in some current flow through R_2. This current flow must be added to the base current in calculating the value of R_1.

Coupling capacitor. When transformer T_1 is used, capacitor C_1 is omitted. When used, C_1 forms a high-pass filter with R_2, producing some loss at the low-frequency end of the audio range. Using a 1-dB loss as the low-frequency cutoff point, the value of C_1 can be found by

$$\text{Capacitance (F)} = \frac{1}{3.2 \times \text{frequency (Hz)} \times \text{resistance (Ω)}}$$

5-10. DESIGNING LOW-POWER AUDIO AMPLIFIERS USING PLASTIC TRANSISTORS

The use of plastic transistors in the output circuits of low-power audio amplifiers permits the designer to get maximum performance at minimum component cost. Motorola Semiconductor Products Inc. has developed several such transistors, together with *complementary-symmetry* audio amplifier circuits which take full advantage of the transistor characteristics. In this section, we shall discuss design of three complementary-symmetry circuits, together with power supplies for the circuits.

The circuits can be fabricated using the component values and supply voltages shown in the parts list of the related schematic diagrams. However, the circuits are quite flexible, and can serve as a starting point for design of similar circuits. To make full use of this flexibility, the circuit diagrams and parts lists are supplemented with design notes which explain the why and how of circuit performance.

Sec. 5-10 Designing Low-Power Audio Amplifiers Using Plastic Transistors 157

5-10.1. Three-Transistor Audio Amplifier

Portable and toy phonographs that use ceramic cartridges of 1- to 3-V output do not usually require low distortion, extended frequency response, or high-input sensitivity. Using a simple complementary-symmetry circuit in these applications can result in significant cost savings over other amplifiers. The basic three-transistor circuit, shown in Fig. 5-31, can provide 2-W output with suitable component and power supply values.

Table 1 in Fig. 5-31 shows the performance of the various versions of the circuit. Table 2 in Fig. 5-31 lists the output power of each version operated at the higher impedance loads. The following design notes apply to all versions of the Fig. 5-31 circuit.

Supply voltage. The supply voltage (V_{CC}) is chosen to provide the necessary signal swing across the speaker. V_{CC} must also be of sufficient magnitude to offset the losses in the transistors and associated components.

The equation for the required V_{CC} is

$$V_{CC} \approx 2\sqrt{(P_{OUT(rms)}R_L)} + V_{CE(sat)}Q_2 + V_{BE(sat)}Q_3$$
$$+ V_{CE(sat)}Q_1 + I_{EQ_1}(R_4)2I_{OUT\ peak}(R_7) \quad (5\text{-}1)$$

The first term in the equation gives the voltage across the load; the remaining terms give the added voltage required to make up for the saturation losses. The sum of the second term, $V_{CE(sat)}Q_2$, and one-half of the last term is the saturation loss on the positive half-cycle, and the remaining terms are the losses on the negative half-cycle.

Center voltage. The center voltage (or the voltage at the junction of emitter resistors R_7 and R_8) must be *one-half of the supply voltage* to ensure maximum signal swing at the speaker. The equivalent circuit in Fig. 5-32 shows the d-c feedback mechanism which maintains the center voltage.

Equations for the center voltage are

$$V_{center} = V_{CC} - (I_C Q_1)(R_6) \quad (5\text{-}2)$$
$$V_{center} = V_{CC} - (K I_B Q_1)(R_2 + R_3) \quad (5\text{-}3)$$

The constant-K is a factor of 5 or greater, which allows sufficient current in the R_2-R_3 divider to prevent its being loaded by the base current of Q_1.

The voltage at the base of Q_1 is slightly greater than the base-emitter voltage of Q_1. The difference appears as a voltage drop across R_4, which sets the collector current of Q_1. This determines the center voltage, according to Eq. (5-2). If the center voltage increases, the voltage at R_4 increases, the collector current of Q_1 increases, the voltage drop across R_6 increases, and the center voltage decreases to a value approximating the original.

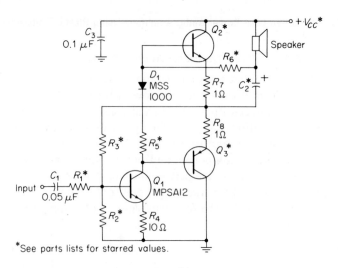

*See parts lists for starred values.

Parts List

Part No.	1 W, 8 Ω	1 W, 16 Ω	1 W, 40 Ω	2 W, 16 Ω	2 W, 8 Ω
V_{CC}	12 V	14 V	20 V	18 V	15 V
R_1	560 kΩ	680 kΩ	820 kΩ	560 kΩ	680 kΩ
R_2	560 kΩ	560 kΩ	750 kΩ	470 kΩ	560 kΩ
R_3	2.2 MΩ	2.7 MΩ	5.6 MΩ	3.3 MΩ	2.7 MΩ
R_5	27 Ω	27 Ω	33 Ω	22 Ω	12 Ω
R_6	560 Ω	820 Ω	1.0 kΩ	220 Ω	200 Ω
C_2	250 μF	150 μF	100 μF	150 μF	250 μF
Q_2	MPS6560	MPS6561	MPSA05	MPS6560	MPSU01*
Q_3	MPS6562	MPS6563	MPSA55	MPS6562	MPSU51*

*Uniwatt

Table 1: Performance of the Three-Transistor Amplifier

Performance Feature	1 W, 8 Ω	1 W, 16 Ω	1 W, 40 Ω	2 W, 16 Ω	2 W, 8 Ω
Sensitivity at rated P_{OUT}	0.8 V	1.15 V	0.82 V	1.2 V	1.4 V
Input current, rated P_{OUT}	160 mA	110 mA	70 mA	180 mA	240 mA
Idling current	9.0 mA	8.0 mA	9.0 mA	10 mA	17 mA
Input impedance	560 kΩ	680 kΩ	820 kΩ	560 kΩ	680 kΩ
Approximate high-frequency cutoff	20 kHz	13 kHz	10 kHz	14 kHz	13 kHz

Table 2: Performance of Three- and Four-Transistor Circuits with Different Impedances

Amplifier Version	P_{OUT}	16-Ω load Average d-c Current	P_{OUT}	40-Ω load Average d-c Current
1 W, 8 Ω	0.8 W	100 mA	0.35 W	60 mA
1 W, 16 Ω	1.0 W	110 mA	0.65 W	70 mA
2 W, 8 Ω	1.0 W	110 mA	0.65 W	70 mA
2 W, 16 Ω	2.0 W	180 mA	1.1 W	90 mA

Figure 5-31 Three-transistor audio-amplifier circuit using plastic transistors (Courtesy Motorola)

Sec. 5-10 Designing Low-Power Audio Amplifiers Using Plastic Transistors 159

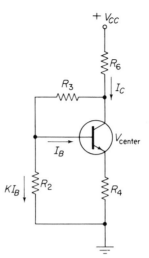

Figure 5-32 Equivalent circuit for three-transistor amplifier showing the feedback path for stabilizing I_C and V_{center} (Courtesy Motorola)

Closed-loop gain. The closed-loop a-c gain in the Fig. 5-31 circuit is set by the ratio of R_3 to R_1. The d-c conditions of the circuit determine the value of R_3. The input impedance of the circuit is approximately equal to the resistance of R_1.

Crossover distortion. To offset crossover distortion, which is most evident at low signal levels, the output transistors are forward biased to allow a small "idling" current through their collectors (refer to Sec. 5-3.3).

In the Fig. 5-31 circuit, the idling current is produced by the forward voltage of D_1 and the voltage drop across R_5. However, heating of the output transistors by signal power and elevated ambient temperatures could cause this idling current to increase to a value which might cause *thermal runaway*. For those not familiar with thermal runaway, a brief description is in order. When current passes through a transistor junction, heat will be generated. If all of this heat is not dissipated by the case, the junction temperature will increase. This, in turn, will cause more current to flow through the junction even though the voltage, circuit values, etc., remain the same. In turn, this will cause the junction temperature to increase even further, with a corresponding increase in current flow. If the heat is not dissipated by some means, the transistor will burn out and be destroyed.

In the Fig. 5-31 circuit, the thermal runaway problem is prevented by the compensating negative temperature coefficient of bias diode D_1, which reduces the idling current to a safe value. Diode D_1 is mounted near one output transistor so that the base-emitter junction and diode are the same temperature. Thus, the voltage drops across diode D_1 and the base-emitter junction are the same and remain the same with changes in temperature.

As an example, should temperature rise, the emitter and collector currents of the transistor also tend to rise. The same temperature rise causes the forward resistance of D_1 to decrease. As a result, the current flowing through R_6 increases, causing an increase in the voltage drop across R_6. Since this voltage drop tends to reverse-bias the emitter-base junction, the net forward bias for that junction is reduced, and the emitter-collector currents are reduced to normal (or tend to reduce to normal). Should the temperature fall, the action is reversed, and the emitter-collector currents are raised toward their normal values.

It is recommended that the 1-W 8-Ω, as well as the 1-W and 2-W 16-Ω circuits use the heat sink shown in Fig. 5-33 for one output transistor and D_1. The extra heating induced in the diode by the power dissipated in the output transistor will cause an even greater reduction in idling current. The other output transistor should be mounted in a heat sink similar to the Staver F1-7.

The combination of the diode and heat sink allows safe operation to about 60°C, while keeping crossover distortion at a minimum.

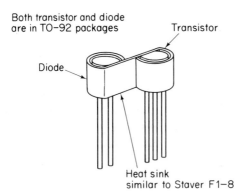

Figure 5-33 Bias diode and one output transistor mounted on the same heat sink (Courtesy Motorola)

a-c response. The a-c response of the Fig. 5-31 circuit is determined by capacitors C_1 and C_2, and the input capacitance of Q_1. Transistor Q_1 is of the Darlington type which has very high gain. The input capacitance is the Miller capacitance (collector-to-base), plus the base-emitter capacitance.

The high-frequency cutoff point is determined by

$$\text{High-frequency cutoff} = \frac{1}{6.28F \times R_{IN} \times C_{IN}} \quad (5\text{-}4)$$

where R_{IN} is the resistance of R_1, and C_{IN} is the input capacitance of Q_1. Typically, the high-frequency cutoff is between 10 kHz and 20 kHz, depending on which version of the circuit is used.

The low-frequency cutoff point is determined by the input impedance in series with C_1 and the speaker impedance in series with C_2. An approximate low-frequency cutoff point can be found using the equations of Sec. 5-1.

Sec. 5-10 Designing Low-Power Audio Amplifiers Using Plastic Transistors   161

Bypass capacitor C_3 is connected from V_{CC} to ground to prevent possible oscillation caused by high-current transients in the power supply leads.

Grounding the loudspeaker. In some systems, it may be desirable to ground the "cold" side of the speaker, instead of connecting the speaker to the supply voltage. This is accomplished by using the alternate circuit of Fig. 5-34. The positive half-cycle drive current for Q_2 is furnished through R_6. Assuming that C_4 is absent, it is apparent that the voltage drop across R_6 is going to be reduced as the output is driven positive, thus reducing the drive current to Q_2. This causes clipping at a point much lower than V_{CC}.

In the basic circuit of Fig. 5-31, output capacitor C_2 is charged to one-half of V_{CC}. When the output voltage rises to V_{CC}, the voltage on the load side of C_2 is effectively 1.5 times V_{CC}. This "extra" d-c charge allows full drive current to Q_2 on the positive excursion of the output. In the alternate circuit of Fig. 5-34, the bootstrap capacitor C_4 provides the same effect.

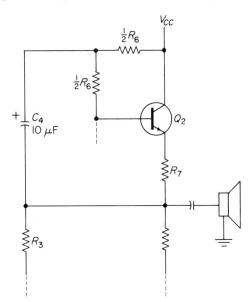

Figure 5-34 Alternate circuit for grounding the loudspeaker (Courtesy Motorola)

5-10.2. Four-Transistor Audio Amplifier

Many applications require low-power audio amplifiers with fairly high-input sensitivity, high-input impedance, and low distortion. Examples are AM and AM/FM radios, radio-phonographs, and small console-type phonographs which use low-output crystal cartridges. The four-transistor complementary-symmetry circuit shown in Fig. 5-35 is well suited for such applications.

Table 1 in Fig. 5-35 shows the sensitivity of the various versions of the

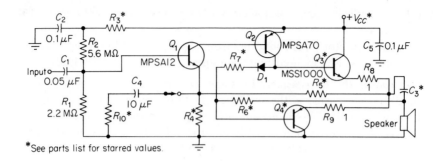

*See parts list for starred values.

Parts List

Part No.	1 W, 8 Ω	1 W, 16 Ω	1 W, 40 Ω	2 W, 8 Ω	2 W, 16 Ω
V_{CC}	12 V	14 V	20 V	14 V	18 V
R_3	3.3 MΩ	5.6 MΩ	6.8 MΩ	7.5 MΩ	5.6 MΩ
R_4	1.0 kΩ	1.0 kΩ	1.2 kΩ	1.0 kΩ	1.0 kΩ
R_5	3.3 kΩ	3.9 kΩ	3.9 kΩ	3.9 kΩ	3.9 kΩ
R_6	560 Ω	560 Ω	1.0 kΩ	820 Ω	390 Ω
R_7	27 Ω	27 Ω	27 Ω	18 Ω	18 Ω
R_{10}	Adjust for desired sensitivity; 200 Ω minimum for 1 MΩ Z in.				
C_3	250 μF	150 μF	100 μF	150 μF	250 μF
Q_3	MPS6560	MPS6561	MPSA05	MPSU01	MPS6560
Q_4	MPS6562	MPS6563	MPSA55	MPSU51	MPS6562

Table 1: Sensitivity of the Four-Transistor Amplifiers with $C_4 = 10$ μF and $R_{10} = 200$ Ω

Amplifier	Sensitivity with C_4 and R_{10}	Sensitivity without C_4 and R_{10}
1 W, 8 Ω	200 mV	600 mV
1 W, 16 Ω	185 mV	800 mV
1 W, 40 Ω	260 mV	1350 mV
2 W, 16 Ω	230 mV	900 mV
2 W, 8 Ω	200 mV	1100 mV

Table 2: Performance of the Four-Transistor Amplifiers

Performance Feature	1 W, 8 Ω	1 W, 16 Ω	1 W, 40 Ω	2 W, 8 Ω	2 W, 16 Ω
Sensitivity at rated P_{OUT}	0.6 Vrms	0.8 Vrms	1.35 Vrms	0.9 Vrms	1.1 Vrms
Input current at rated P_{OUT}, 10% THD	180 mA	110 mA	82 mA	240 mA	145 mA
Quiescent idle current	12 mA	13 mA	13 mA	16 mA	15 mA

Total harmonic distortion: 10% at rated P_{OUT}
  2% at ½ rated P_{OUT}, measured with regulated power supply

Figure 5-35 Four-transistor audio-amplifier circuit using plastic transistors (Courtesy Motorola)

Sec. 5-10 Designing Low-Power Audio Amplifiers Using Plastic Transistors 163

circuit. Table 2 in Fig. 5-35 lists the performance of the various versions. The following design notes apply to all versions of the Fig. 5-35 circuit.

Center voltage tracking. To ensure maximum undistorted signal swing at the speaker, the center voltage (at the junction of R_8 and R_9) must be one-half of V_{CC}. For low distortion and symmetrical clipping with variations in V_{CC}, the center voltage should "track" V_{CC}. That is, when V_{CC} changes, the center voltage should change by a corresponding amount. This tracking should be independent of variations in transistor characteristics.

Center voltage tracking is accomplished by the d-c feedback characteristics of the amplifier. The voltage at the base of Q_1 is set by the divider action of R_1, R_2 and R_3, thus setting the voltage at the top of R_4. The currents through R_4 and R_5 are almost the same and are approximately equal to the emitter voltage of Q_1 divided by the resistance of R_4. The current through R_4 and R_5 is much greater than the emitter current of Q_1, so variations of the emitter current caused by changes in Q_1 gain will have little effect on the divider current.

Transistor Q_2 is biased as a class A amplifier (refer to Sec. 5-3.1). The d-c load resistor for Q_2 is R_6. Transistors Q_3 and Q_4 are slightly forward-biased (almost class AB, see Sec. 5-3.3), so the collector of Q_2 and the top of R_6 are at almost the same d-c point as the top of R_5.

If the center voltage was to decrease, the voltage at the top of R_4 would also decrease. Since the voltage at the base of Q_1 is set by the resistive bias network (R_1, R_2, R_3) and is constant, a decrease in the voltage at the emitter of Q_1 will cause an increase in collector current of Q_1. Since the collector current of Q_1 is the base current of Q_2, an increase in this current will cause the Q_2 collector current to increase, raising the voltage drop across R_6 and increasing the center voltage toward its original value.

If V_{CC} drops, due to loading or changing line voltage, the voltage at the base of Q_1 also drops. This decreases the drop across R_4 and lowers the center voltage. In a similar fashion, the circuit of Fig. 5-35 will compensate for changes in center voltage or V_{CC} due to loading, supply voltage, or changes in transistor characteristics.

Closed-loop gain. The d-c closed-loop gain of the Fig. 5-35 amplifier circuit is set by the ratio: $(R_5 + R_4)/R_4$, and is quite low, allowing maximum d-c stability. The a-c closed-loop gain is also determined by this ratio, but may be increased considerably by adding C_4 and R_{10}. The a-c gain is then approximately equal to R_5 divided by R_{10}. Table 1 in Fig. 5-35 shows the sensitivity for the various versions of the circuit without C_4 and R_{10}, and with R_{10} equal to 200 Ω and C_4 equal to 10 µF.

Crossover distortion. To prevent crossover distortion, the output transistors are slightly forward-biased by the forward voltage drop of D_1 and the drop across R_7.

On the negative half-cycle of the signal, drive current for output transistor Q_4 is through R_6. To prevent the loss of drive to Q_4 which would occur on large negative signals, resistor R_6 is connected to the loudspeaker. This allows the output coupling capacitor C_3 to act as a bootstrap capacitor and supply adequate drive current to Q_4.

Filtering. Capacitor C_5 filters the supply to prevent possible oscillation caused by high-current surges. The base of Q_1 is biased by the supply voltage. It is necessary to filter this d-c bias voltage to prevent powerline hum and noise being amplified by the circuit. Resistor R_3 and capacitor C_2 form a high-pass filter with a 3-dB rolloff point at about 1 Hz. This effectively filters the bias voltage at the base of Q_1.

Frequency response. Low-frequency response of the amplifier is determined by C_1 and C_3. Typically, cutoff is about 50 Hz. The high-frequency response is determined essentially by the transistor characteristics and is typically greater than 20 kHz.

Input impedance. The circuit of Fig. 5-35 offers a high-input impedance. Use of a Darlington amplifying transistor for Q_1 ensures that the input impedance will be primarily determined by the input biasing resistors R_1 and R_2. Typically, the input impedance is greater than 1 MΩ for any of the Fig. 5-35 circuit versions.

Heat sinking. Heat sinking of the output transistors is required for reliable operation at high ambient temperature and high line voltage. All comments on heat sinking for the three-transistor amplifier circuit (Sec. 5-10.1) apply to the four-transistor circuit. With these heat sinking arrangements (diode on common heat sink with one output transistor, and the other output transistor also heat sinked), reliable operation to about 60°C is possible.

5-10.3. Power Supplies for Amplifier Circuits

The continuous output power of the amplifiers is dependent upon regulation of the power supply. The usual procedure for phonograph amplifiers is to use an overwinding on the motor as the secondary of the power transformer. AM/FM radios usually use an inexpensive transformer which has relatively poor regulation. Either of these methods results in a drop of 10 to 25 per cent in supplied voltage at maximum required output current. Basically, this means that the average 1-W or 2-W amplifier is not capable of sustaining these power levels continuously.

As an example, Fig. 5-36 shows a schematic of half-wave and full-wave power supplies with a chart of voltage versus load resistance. The output

Sec. 5-10 Designing Low-Power Audio Amplifiers Using Plastic Transistors 165

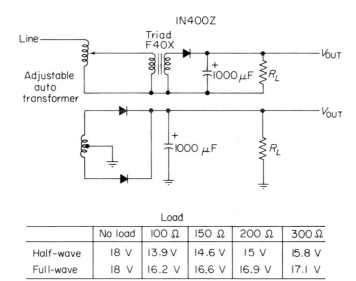

	No load	100 Ω	150 Ω	200 Ω	300 Ω
Half-wave	18 V	13.9 V	14.6 V	15 V	15.8 V
Full-wave	18 V	16.2 V	16.6 V	16.9 V	17.1 V

Figure 5-36 Power supply voltage regulation with load variations (Courtesy Motorola)

voltage of the half-wave circuit drops from 18 V to 13.9 V with a 100-Ω load. This is about the same load as a 1-W 16-Ω amplifier under loaded conditions. (An 18 V supply is not sufficient for a 2-W 16-Ω amplifier).

In general, if 1 W of continuous power amplification for a 16-Ω load is needed, any power supply that can deliver 2 W of program material is satisfactory. Most musical program material consists of a fairly low-to-average sound level punctuated by occasional high-power peaks. Even the simple half-wave supply shown in Fig. 5-36 is capable of delivering and sustaining the desired power level of music program material.

Effects of filter capacitance. The average output of the power supplies shown in Fig. 5-36 consists of a d-c voltage with some amount of hum (dependent upon the current drawn) superimposed on this voltage. With the full-wave supply and a capacitor of 500 μF, a peak-to-peak hum of 3 V and an average d-c voltage of 14 V is typical. When the capacitance is increased to 1000 μF, as shown, the d-c output voltage increases to about 14.15 V but the hum drops to about 1 V.

Under continuous load conditions, the hum appears in the amplifier output on peak positive signal excursions. With a load consisting of music program material, the hum level is quite low due to the low current drawn from the supply. For this reason, a 500 μF filter capacitor is adequate (at least for music program material).

5-10.4. Line-Operated Amplifier

In some applications, it is desirable to use the full a-c line voltage (110 V to 120 V) as the power source for amplifiers. This eliminates the need for a supply transformer. Typical applications for a line-operated amplifier include a table radio and a portable television set. The line-operated amplifier of Fig. 5-37 meets the needs of such applications. The circuit of Fig. 5-37 has a *single-ended* output, unlike the amplifiers discussed in Secs. 5-10.1 and 5-10.2.

Circuit performance of the amplifier is shown in the table 1 of Fig. 5-37.

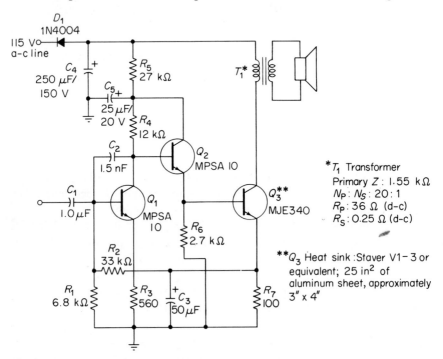

Sensitivity and Output Power at Various Load Impedances

Input Voltage	Output Power	Load Impedance
15 mV	1.8 W	3.2 Ω
15 mV	2.0 W	8.0 Ω
15 mV	2.6 W	16 Ω

Average current drawn : 52 mA

D-C power dissipated in Q_3 : 6 W

*Output power at 10 % THD

Figure 5-37 Line-operated power amplifier using plastic transistors (Courtesy Motorola)

Sec. 5-10 Designing Low-Power Audio Amplifiers Using Plastic Transistors 167

The output power listed in the table was measured with the transformer T_1 shown. Output power of the amplifier is dependent upon T_1 and load impedance. The following design notes apply to all of the load impedances listed in the table.

d-c stability and a-c gain. As with any direct-coupled amplifier (Sec. 5-2), d-c stability is important to a line-operated amplifier. The required stability is provided by d-c feedback from the emitter of Q_3 to the base of Q_1, through R_2. To prevent the loss of a-c gain that could result from a-c feedback, capacitor C_3 is used to bypass the audio part of the feedback.

The overall a-c gain of the amplifier is dependent on the gain of the Q_1 stage (which is fixed by its load and emitter impedances and is fairly low) and the gain of the output stage Q_3 (which is dependent on the primary impedance of transformer T_1). If desired, a resistor may be inserted in series with C_3 to provide some a-c feedback. Of course, this will reduce sensitivity, but it will also make the circuit gain more predictable.

Power supply circuitry. In typical table radios, part of the power amplifier circuitry provides a supply voltage for the IF and converter stages. The voltage at the junction of R_4 and R_5 can be used for this purpose. Resistor R_5 must be reduced to compensate for the added current. Capacitor C_5 filters hum and noise from this supply voltage.

Note that the collector of Q_2 is operated from this supply voltage; this allows the use of a transistor with a low BV_{CEO} (breakdown voltage) for Q_2 instead of a 300-V transistor that would be required if the collector of Q_2 were attached to the collector of Q_3.

Neutralization. High-frequency oscillation of Q_1 may occur with varying source impedances. Capacitor C_2 provides a form of neutralization to offset any tendency for oscillation. Neutralization is discussed further in Chapter 6.

Breakdown protection. In any transformer-coupled amplifier circuit, the output transistor is subject to several possible failure modes. Removal of the load from the secondary could cause a several-hundred-volt transient in the primary, which could exceed the BV_{CEO} rating of the transistor, and cause destructive secondary breakdown.

Figure 5-38 shows three methods of preventing such breakdown. First, a voltage dependent resistor (VDR) can be placed across the primary of the output transformer. Second, a resistor with a value of about 5 to 10 times the load resistance can be placed across the transformer secondary. Third, in television sets where a 250 to 300 V is available (typically from the video amplifier supply), the collector of Q_3 can be clamped to the supply voltage by a diode. Of course, the supply voltage must be less than the BV_{CEO} of Q_3. In general, most failures are caused by exceeding the BV_{CEO} of the output transistor Q_3.

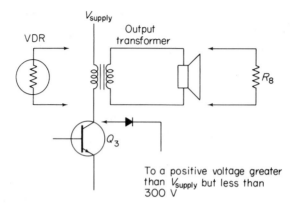

Figure 5-38   Protective methods for line-operated amplifier (Courtesy Motorola)

Power dissipation. The other major cause of device failure and excessive power dissipation is generally due to inadequate heat sinking or excessive ambient temperature. The output transistor is operated as class A, and is dissipating about 6 W. For safe operation to about 50°C ambient, a heat sink with a minimum thermal resistance of 10°C/W is required. A heat sink such as the Staver V1-3, or an aluminum sheet having about 25 in² of radiating area should be used.

Power supply circuit. Power for the amplifier circuit is provided by diode D_1, a type IN4004 Motorola Surmetic rectifier, and C_4, a 250 μF 150-V capacitor. A capacitor with lower capacitance may be used if desired. However, output power will be reduced due to the resultant decrease of supply voltage.

5-10.5. Miscellaneous Considerations

The amplifiers in Figs. 5-31 and 5-35 are very flexible. Different power levels and load impedances can be obtained by making minor changes in the bias networks and by specifying the transistors to match the application.

For example, if a ½-W amplifier for a 16-Ω load is desired, the 1-W 8-Ω version of either circuit can be used. The output transistors recommended for the 1-W 8-Ω amplifier are 500-mA devices. Thus, a cost saving can be realized by using a lower-current device such as the Motorola MPSA05 and MPSA55.

In applications that do not require high-input impedance, a low-cost NPN transistor such as the Motorola MPSA10 can be substituted for the MPSA12 Darlington amplifier. The bias resistors would then have to be reduced by approximately *one hundred,* and the ratio of resistance altered to

Sec. 5-10 Designing Low-Power Audio Amplifiers Using Plastic Transistors 169

compensate for the $V_{BE(ON)}$ (base-emitter turn-on voltage) of the single transistor.

If a different power level is desired, it is only necessary to compute the required V_{CC}, and then adjust the bias resistors to obtain one-half of V_{CC} at the center point. The output transistors should then be specified, taking into consideration the peak curent, maximum V_{CC} (allowing 20 per cent over for high line and transformer tolerances), and maximum power dissipation.

6 RF AMPLIFIER
DESIGN EXAMPLES

Both two-junction transistors and FETs are used as RF amplifiers. Generally, FETs are limited to voltage amplifier amplifications. Two-junction transistors can be used in either voltage amplifier or power amplifier circuits. Since the number of different amplifier circuits is almost unlimited, it is impossible to cover all aspects of RF amplifiers in this chapter. Instead, we shall concentrate on a *cross section* of RF amplifier design examples.

As discussed in Chapter 5, the author's *Handbook of Modern Solid-state Amplifiers* (Prentice-Hall, Inc., Englewood Cliffs, N.J., 1974) provides a detailed discussion of many amplifier types.

6-1. RF VOLTAGE-AMPLIFIER DESIGN REQUIREMENTS

Radio-frequency voltage amplifiers are used primarily in receivers and receiver-type circuits. An IF (intermediate frequency) amplifier, or IF limiter amplifier, is an example of an RF voltage amplifier. The input, or first stage, of a receiver may include a separate RF voltage amplifier (such as with some communications receivers). However, most solid-state receivers combine the RF voltage-amplifier function with that of the local oscillator. Such circuits are discussed in Sec. 6-2.

Figure 6-1 shows the working schematic of a typical RF voltage amplifier. Such a circuit could be used as an IF amplifier, IF limiter, or a separate RF amplifier with few modifications. Both the input and output are tuned to the desired operating frequency by means of the resonant circuits. In this case, the resonant circuits are composed of transformers with a capacitor across

Sec. 6-1   RF Voltage-Amplifier Design Requirements   171

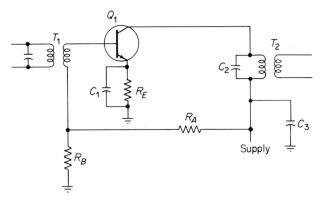

Voltage gain $\approx \beta \left(\frac{1}{N}\right)$
Where N is turns ratio of T_2

$N \approx \sqrt{\frac{Z_P}{Z_S}}$ $N^2 = Z_P/Z_S$
Where $Z_P = T_2$ primary impedence
      $Z_S = T_2$ secondary impedence

Supply $\approx$ 3 to 4 times desired output voltage
Voltage drop across $R_A \approx$ supply − drop across R_B; $R_B \approx 10 R_E$
Voltage drop across $R_E \approx$ 0.2 for germanium, 0.5 for silicon
At operating frequency: $XC_1 \lesssim Z_{IN}$ of Q_1; $XC_2 \approx$ primary of T_2; $XC_3 \lesssim 100\ \Omega$

Figure 6-1   Basic RF voltage-amplifier design

the primary. The capacitors can be variable, but are usually fixed. The resonant circuit is tuned by an adjusting slug between the windings.

6-1.1. Circuit Analysis

The considerations for transformer-coupled RF amplifiers are similar to those of audio amplifiers, as described in Chapter 5. However, the guidelines for trial values are somewhat different.

Transistor characteristics. The overall considerations for selection of transistors in any solid-state circuit apply to RF amplifiers. Of particular importance are interpreting datasheets and determining parameters at different frequencies. The temperature-related problems generally do not apply, since RF voltage amplifiers usually operate at very low-power levels.

The main concern is that the transistor provide the required gain at the frequency of interest. In general, the transistor should provide 1.5 times the required gain at the operating frequency. This will compensate for mismatch variation in gain due to differences in transistors, etc.

Transformer characteristics. It is often practical to design an RF amplifier around the characteristies of a commerical transformer (interstage IF transformer, IF detector transformer, RF/IF transformer, etc.). Such transformers are usually rated as to primary and secondary impedance (rather than turns ratio), and possibly current capacity. However, the typical low currents involved present no design problems.

Some commercial transformers are provided with a built-in fixed capacitor across the primary (and/or secondary in some cases). When a capacitor is used, the transformers are rated as to the resonant frequency range or midpoint (455 kHz for an AM broadcast IF; 500 to 1600 kHz for an RF input transformer, often of the ferrite "loopstick" type; 10.7 MHz for an FM broadcast IF; etc.). In other transformers, a fixed or variable capacitor must be connected across the transformer windings. For example, a "loopstick" requires a variable capacitor of the given range to provide full tuning across the AM broadcast band.

Stage gain. Voltage gain of a fully bypassed RF amplifier is approximately equal to transistor gain (β) and the turns ratio of the output transformer, as shown in Fig. 6-1. When the turns ratio is considered as primary/secondary, the stage gain equals transistor β times the *inverse* of the turns ratio. For example, if the transistor gain is 10, and the transformer has a turns ratio of 10 (primary) to 1 (secondary), the net voltage gain is 1 (or unity).

The required stage gain depends upon the circuit application. As guidelines, a communications receiver RF stage requires a gain of between 10 and 20, an AM broadcast IF stage requires a gain of between 30 and 40, and an IF amplifier for FM requires a gain between 40 and 50; a television IF amplifier (broadband) requires a gain between 15 and 20.

Supply voltage. The value of supply voltage for an RF voltage amplifier is not critical. Of course, the supply voltage should be between three and four times the desired output voltage for the stage. In most receiver RF circuits, the desired output in between 1 and 2 V, so the supply could be between 3 and 9 V. A higher supply voltage can be used, provided the transistor characteristics are not exceeded.

Emitter resistance. When the emitter resistance R_E is bypassed, the resistance value should be chosen on the basis of direct current, rather than signal. The value of R_E should provide a voltage drop equal to the emitter-base voltage differential when the normal (d-c operating point) collector current is flowing. A typical silicon emitter-base differential is about 0.5 V (0.2 V for germanium transistors). The drop across R_E will serve to stabilize the gain of Q_1.

To get a perfect impedance match between transistor and output transformer (a practical impossibility), the total impedance presented by R_E and the transistor should match the transformer primary impedance. As a guideline, assume that the impedance represented by the full collector voltage and current (V_C/I_C) is the total transistor and R_E impedance. This will establish the desired collector current I_C and a corresponding voltage drop across R_E. For example, assume that the supply voltage is 10 V, and the primary impedance is 10 kΩ. Ignoring the small d-c drop across the primary, the collector is at 10 V. The desired collector current to match impedance is

$$I = \frac{E}{R} = \frac{10}{10,000} = 0.001 \text{ A}$$

or 1 mA. With 1mA flowing through R_E (at the operating point) and a desired 0.5-V drop, the value of R_E is

$$R_E = \frac{E}{I} = \frac{0.5}{0.001} = 500 \text{ Ω}$$

Bypass capacitors. The value of the emitter-bypass capacitor C_1 should be such that the reactance, at the lowest operating frequency, is less than the input impedance of the transistor. This will effectively remove the emitter resistor from the circuit, so far as signal is concerned. The input impedance of a typical two-junction transistor for RF applications is on the order of a few ohms and given on the datasheet. If the input impedance is not known, use a capacitor that will produce a reactance of less than 10 Ω at the operating frequency.

The value of supply line bypass capacitor C_3 is between 0.001 and 0.01 μF in a typical RF voltage amplifier. As a first trial value, use a value for C_3 that will produce a reactance of less than 100 Ω at the operating frequency.

Bias resistance network. The values of the bias resistance network should be chosen to place transistor Q_1 at the desired operating point. For example, if the desired collector current is 1 mA and Q_1 has a nominal gain of 10, the base current must be 0.1 mA. Likewise, if the emitter-base voltage differential is assumed to be 0.5 V, with another 0.5-V drop across R_E, the base should be at 1 V under no-signal conditions. Any combination of R_B and R_A that produces these relationships would be satisfactory. As a first trial value, make R_B ten times the value of R_E. Then calculate a corresponding value for R_A, using the equations in Fig. 6-1.

When the circuit in Fig. 6-1 is to incorporate an AVC-AGC (automatic volume control-automatic gain control) function, the bias network is also used as the AVC-AGC line. Thus, the bias network values must be calculated on that basis. A discussion of AVC-AGC circuits is covered in Sec. 6-3.

6-2. RF MIXERS AND CONVERTERS

It is possible to design RF mixers and converters using the admittance parameters described in Sec. 2-2. Such a design approach is based on the popular two-port parameter design method, also used for voltage amplifiers as discussed in Secs. 2-2 and 6-6.

6-2.1. Design Considerations

In the design of mixers using admittance parameters, the following items are of considerable importance in obtaining good overall circuit performance.

1. Frequencies
2. Stability
3. Gain
4. Network design
5. Local oscillator injection
6. Device selection

Each of these items are important in the mixer circuit design considerations for the following reasons:

Frequencies. Mixing action is done by means of a non-linear device, therefore, many different frequency components are present in the output of the circuit. These frequency components may be classed as:

1. *Spurious mixer products.* All frequency components in the output of the mixer, other than the desired sum or difference output component.
2. *Crossovers or "birdies".* Undesired mixer frequency components that fall within the mixer output passband.
3. *Intermodulation—distortion products.* A special class of spurious mixer products falling within the mixer output passband and resulting from interaction between signal components fed into the mixer.

Spurious output signals are a problem with any mixer design. In addition to the obvious outputs of the local oscillator frequency, the input RF signal frequency, and the undesired sum or different frequency, many other spurious output signals may be present. Many of these additional undesired outputs are due to third and higher order distortion characteristics that the non-linear device exhibits in addition to the second order distortion used by such a device to produce the desired mixing action.

Spurious output signals at both the desired output frequency, and other frequencies may be present. The undesired frequency signals may be attenuated by tuned circuits following the mixer, but the only recourse to spurious outputs at the desired output frequency may be an entirely different mixing scheme.

The information in this section assumes that the designer has already selected suitable mixing frequencies and is ready to proceed with the actual mixer design.

Stability. The stability problem of mixers may be looked at basically from a two-port standpoint. Both the Linvill C and Stern k factors, described in Sec. 2-2, apply to mixer design. However, for the collector circuit of a mixer, the stability factor should be calculated at the output frequency. For the base circuit, the stability factor should be calculated at the input frequency. Note that the greatest danger of oscillation occurs at the output frequency since the impedance level at the collector is higher.

If the output port presents a low impedance or short circuit at the input frequency and the input port presents a low impedance or short circuit at the output frequency, then oscillation should not occur.

In case it is impossible to meet these stability conditions, as calculated from the Stern k factor, it is still possible to proceed with mixer design using the same transistor. This requires external tuning networks. For example, an IF trap can be used at the input of the two-port network, and an RF trap can be used at the output of the two-port network. Of course, this involves the use of extra components. In general, it is better to select a transistor that will produce the desired stability factor (a Stern k of about three or higher) with the existing source and load.

Gain. Conversion gain is defined as

$$\frac{\text{IF power out}}{\text{RF power in}}$$

Depending upon the frequencies of operation, the mixer conversion gain will vary as does the gain of any amplifier designed for operation at the different frequencies. The gain, however, will be comparable to that obtained in an amplifier designed for the same intermediate frequency (IF) of operation.

Network design. The primary considerations in the design of the input and output networks are: conversion gain, stability, and attenuation of off-frequency spurious output signals.

Local oscillator injection. There are basically two methods for injecting the *LO* (local oscillator) signal, either *base injection* or *emitter injection*. Of these two schemes, base injection usually provides greater stability in the VHF and UHF frequency ranges.

Device selection. Since a transistor is a non-linear device, any transistor can be used as a mixer. However, certain characteristics make some transistors more desirable (as mixers) than others. These characteristics include:

1. Frequency. The device must be capable of operation at the input, output, and local oscillator frequencies. Generally, the local oscillator frequency is the highest of the three. Thus, the LO frequency sets the transistor maximum limits.
2. Gain. The device gain will be within 3 dB when used as a mixer, as opposed to an amplifier designed at the output frequency in the unneutralized condition. Thus, the device must be capable of the desired gain at the output frequency.
3. Stability. It is recommended that a transistor selected for mixer applications have a Stern k factor of at least two, and preferably three or higher.
4. A transistor with a low-input capacity provides for easier impedance matching.

6-2.2. Design Theory

The basic design theory of a mixer circuit is explained here in terms of equivalent circuits with emphasis on impedance-matching techniques of both the input and output of the transistor to the respective networks.

Equivalent circuits. The equivalent circuits of the mixer circuit are shown in Fig. 6-2. This illustration shows mixer with output shorted (Fig. 6-2(a)),

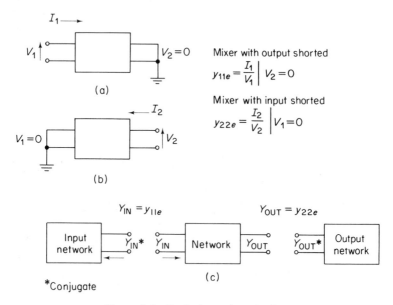

Figure 6-2 Equivalent mixer circuits

Sec. 6-2 RF Mixers and Converters 177

mixer with input shorted (Fig. 6-2(b)) and network interconnections (Fig. 6-2(c)). Note that the terms y_{11e} and y_{22e} in Fig. 6-2 are the common-emitter short-circuit input admittance, and common-emitter short-circuit output admittance, respectively.

The equations and equivalent circuits of Fig. 6-2 illustrate the following. For maximum circuit stability with the input circuit tuned to the radio frequency, the output is considered as a short circuit. Likewise, with the output tuned to the intermediate frequency, the input circuit can be considered a short circuit.

Under these conditions, in the network of Fig. 6-2(c), Y_{IN} represents the input admittance of the transistor, and is equal to the small-signal common emitter input admittance y_{11e} at the input frequency. Y_{OUT} represents the output admittance of the transistor, and is equal to the small-signal common-emitter output admittance y_{22e} at the output frequency.

Conversion gain and conjugate match. For maximum conversion gain the input network should be *conjugately matched* to the transistor input admittance. For example, if the transistor input admittance is $8 + j9$, the network should be designed to produce $8 - j9$. Also, the output network should be conjugately matched to the transistor output admittance. Therefore, in Fig. 6-2(c), Y_{IN}^* is the conjugate admittance of Y_{IN}, and Y_{OUT}^* is the conjugate admittance of Y_{OUT}.

Base injection. In the design of a mixer circuit using base injection for both the input and local oscillator signals, as long as the base does not have sufficient drive from the local oscillator signal to move operation out of the small-signal region, then the small-signal common-emitter short-circuit input admittance will be the design criteria for the source admittance. As the local oscillator signal level is increased, however, the transistor is driven harder into conduction. The input admittance of the transistor changes to a large signal input admittance, this new large signal input admittance then becomes the design criteria for the source admittance.

Assuming operation in the small-signal region and injection of the local oscillator signal at the base (Fig. 6-3) the source must be a short circuit at the output frequency and a y_{11e} conjugate at the input frequency. Likewise, the load should appear as y_{22e} conjugate at the output frequency and a short circuit at the input frequency.

Emitter injection. It is also possible to inject the local oscillator signal at the emitter, as shown in Fig. 6-4. One of the advantages for emitter injection is that there is less chance of the transistor being driven out of the small-signal operating region. However, the disadvantages of emitter injection usually offset the advantages.

With emitter injection, the emitter must still be at RF ground for the input frequency, but at a different impedance level for the local oscillator signal.

This requires a complex network. This network must present a short circuit to the emitter at the input frequency, and present an impedance which is the conjugate of the local oscillator source impedance at the local oscillator fre-

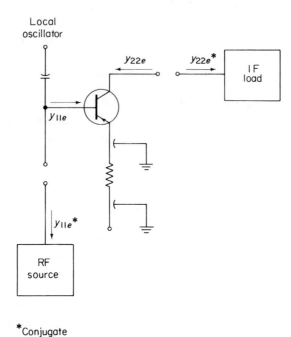

*Conjugate

Figure 6-3 Base injection of both RF and local oscillator signals

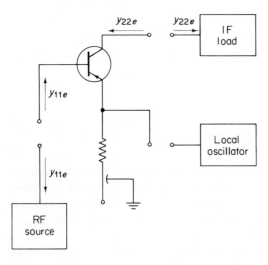

*Conjugate

Figure 6-4 Base injection of RF signal and emitter injection of local oscillator signal

quency. This is a very difficult design problem if the local oscillator and input frequencies are very close together, and are in the VHF or UHF frequency range.

6-2.3. Example of Mixer Circuit Design

The following paragraphs described provide an example of how the design techniques and requirements discussed thus far can be applied to a practical mixer. The mixer circuit, together with some equivalent circuits, is shown in Fig. 6-5. The mixer is used to convert a 30-MHz RF signal to a 5-MHz IF signal with a 35-MHz local oscillator signal injected at the base, using a Motorola 2N2221A transistor.

Transistor admittance parameters. The admittance parameters for the 2N2221A are

$$y_{11e} = 6.25 + j9.5 \text{ mmhos (at 30 MHz)}$$

$$y_{22e} = 0.027 + j0.28 \text{ mmhos (at 5 MHz)}$$

Both admittance values are obtained from the datasheet, or from actual test, using an I_C of 2 mA and a V_{CE} of 10 V.

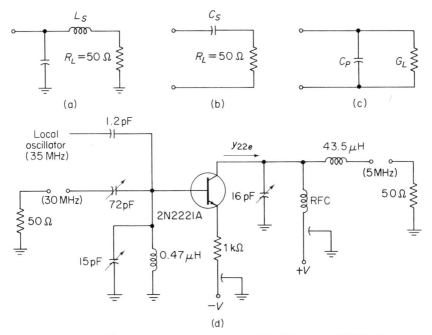

Figure 6-5 Mixer circuit for conversion of 30-MHz RF to 5-MHz IF, using a 35-MHz local oscillator signal (Courtesy Motorola)

Network conjugates. The input network must be the conjugate of y_{11e}, $1/y_{11e}$. The resistance and capacitance values that will produce such a conjugate are found as follows:

The resistance is found by dividing the real part of y_{11e}, 6.25 mmho, into 1. The result is 160 Ω.

The capacitance requires two steps. First, divide the imaginary part of y_{11e}, $j9.5$ mmho into 1 to find a reactance of 105 Ω. Then substitute 105 Ω of capacitive reactance into the conventional equation $1/(6.28FX_C)$ to find a capacitance of 50.5 pF.

The output network must be the conjugate of y_{22e}, $1/y_{22e}$. The resistance and capacitance values that will produce such a conjugate are

$$3700 \text{ Ω and } 9 \text{ pF}$$

Output network circuit. Assume that the output feeds into an IF amplifier, and that the amplifier presents a load of 50 Ω to the output. Thus, the mixer output must be matched to a 50-Ω load. Also assume that the network is to be of the low-pass filter type (to attenuate higher frequencies), such as the one shown in Fig. 6-5(a).

Output circuit Q. The Q of the series inductance in the Fig. 6-5(a) network is found by

$$Q_s = \sqrt{\frac{G_P}{R} - 1}$$

where Q_s is the series Q of the inductance, G_P is the real part of the parallel admittance of the transistor, and R is the load resistance.

Substituting the values,

$$Q_s = \sqrt{\frac{3700}{50} - 1} = 27.3$$

Required series inductance. The series inductance L_S value that will produce the required Q at the frequency and load is found by

$$L_S = \frac{Q_s R}{6.28 F}$$

Substituting the values

$$L_S = \frac{27.3 \times 50}{(5 \times 10^6)(6.28)} = 43.5 \text{ μH}$$

Required parallel inductance. The parallel inductance L_P value that will produce the required Q at the frequency and load is found by

$$L_P = L_S \left(1 + \frac{1}{Q_s^2}\right)$$

For practical purposes, with the value of Q_s, L_P is approximately equal to L_S, or about 43.5 μH.

Required total capacitance. The total capacitance C_T value that will produce resonance at the frequency is found by

$$C_T \text{ (in pF)} = \frac{2.54 \times 10^4}{F \text{ (in MHz)}^2 \times L_S \text{ (in } \mu\text{H)}}$$

Substituting the values,

$$C_T = \frac{2.54 \times 10^4}{(5)^2 \times 43.5} \approx 23.3 \text{ pF}$$

Tuning capacitance value. The value of the output tuning capacitance, when added to the transistor output capacitance, must equal the total capacitance C_T. With a C_T of approximately 23.3 pF, and an output capacitance of 9 pF, the tuning capacitance is $23.3 - 9 = 14.3$ pF. A 16-pF capacitor is selected for the working circuit, as shown in Fig. 6-5(d).

Input network circuit. Assume that the input is from a 50 ohm antenna. Thus, the mixer input must be matched to a 50-Ω load. The series and parallel equivalent input networks are shown in Figs. 6-5(b) and 6-5(c), respectively.

Input circuit Q. The Q of the series capacitance in the Fig. 6-5(b) network is found by

$$Q_s = \sqrt{\frac{G_P}{R} - 1}$$

where Q_s is the series Q of the capacitance, G_P is the real part of the parallel admittance of the transistor, and R is the load resistance.

Substituting the values,

$$Q_s = \sqrt{\frac{160}{50} - 1} = 1.48$$

Required series capacitance. The series capacitance C_S value that will produce the required Q at the frequency and load is found by

$$C_S = \frac{1}{6.28 F Q_s R}$$

Substituting the values,

$$C_S = \frac{1}{6.28(30 \times 10^6)1.48 \times 50} \approx 72 \text{ pF}$$

Required parallel capacitance. The parallel capacitance C_P value that will produce the required Q at the frequency and load is found by

$$C_P = C_S \left(\frac{1}{1 + \frac{1}{Q_s^2}} \right)$$

Substituting the values,

$$C_P = 72 \left(\frac{1}{1 + \frac{1}{1.48^2}} \right) \approx 49 \text{ pF}$$

Required tuning inductance value. The inductance that will produce resonance at the frequency is found by

$$L \text{ (in } \mu\text{H)} = \frac{2.54 \times 10^4}{F \text{ (in MHz)}^2 \times C_P \text{ (in pF)}}$$

Substituting the values,

$$L = \frac{2.54 \times 10^4}{(30)^2 \times 49} \approx 0.6 \ \mu\text{H}$$

In the practical circuit of Fig. 6-5(d), a value of 0.47 μH is used instead of 0.6 μH. This permits an adjustable tuning capacitor (15 pF) to be placed in parallel with the inductance. The 1.2-pF capacitor provides for local oscillator injection. The 1.2-pF value will not significantly affect the 72-pF capacitor (which is also adjustable).

Conversion gain. Figure 6-6 gives conversion gain as a function of local oscillator input levels for different collector currents. Keep in mind that the

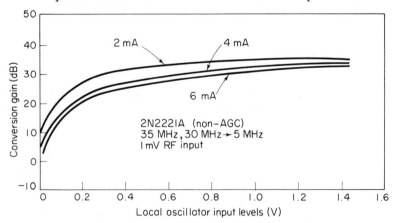

Figure 6-6 Conversion gain of mixer circuit at various levels of collector current (Courtesy Motorola)

y-parameters are a *function of collector current*. Thus, a new set of parameters must be used to design the matching networks for different collector currents.

6-3. AVC-AGC CIRCUITS FOR AMPLIFIERS

Most receivers have some form of AVC-AGC (automatic volume control-automatic gain control) circuit. The terms AVC and AGC are used interchangeably. AGC is a more accurate term since the circuits involved control the gain of an IF or RF stage (or several stages simultaneously), rather than volume of an audio signal in an AF stage. However, in a broadcast receiver the net result is an automatic control of volume. Either way, the circuit functions to provide a constant output despite variations in signal strength. An increased signal will reduce stage gain, and vice versa.

Figure 6-7 shows the working schematic of two AGC systems that are common to broadcast and communications receivers. Diode CR_1 acts as a variable shunt resistance across the input of the IF stages. Diode CR_2 functions as the detector and AGC bias source.

Under no-signal conditions, or in the presence of a weak signal, diode

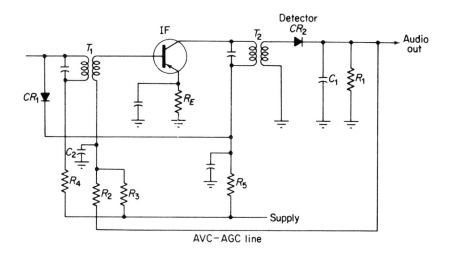

CR_1 is reverse biased with no signal
$C_2 \approx 10 \mu F$
Drop across $R_1 \approx 0.5 - 1.0$ V
Drop across $R_1 + R_2 \approx 1.0 - 2.0$ V
Drop across $R_3 =$ supply $- (R_1 + R_2)$
$R_1 + R_2 \approx 10 \times R_E$

Figure 6-7 Basic AVC-AGC circuit

CR_1 is reverse-biased and has no effect on the circuit. In the presence of a very large signal, CR_1 is forward-biased and acts as a shunt resistance to reduce gain.

The output of CR_2 is developed across R_1 and applied to the audio stages. Resistor R_1 also forms part of the bias network for the IF stage transistor. The combined fixed bias (from the network) and variable bias (from the detector) is applied to the IF stage base-emitter circuit. The detector bias varies with signal strength, and is of a polarity that opposes variations in signal. That is, if the signal increases, the detector bias will reverse bias the transistor.

6-3.1. Design Requirements

Both AGC systems (shunt diode and variable bias) are often found in the same receiver. The variable-bias system handles normal variations in signal. The shunt diode handles large-signal variations.

Shunt diode. The shunt diode CR_1 should have a maximum reverse (peak inverse) voltage rating equal to the supply voltage. In most cases, the diode will have a reverse voltage greater than 1 or 2 V. However, if the diode is capable of handling the full supply voltage, there will be no danger of breakdown. The forward-current capability of CR_1 should be such that the diode can pass the current if there is a full voltage drop across the collector resistors. The values of R_4 and R_5 must be such that CR_1 is reverse-biased under no-signal conditions (with the IF stages at the Q point).

Bias network. The values of the bias network (R_1, R_2, and R_3 should be chosen to provide the desired fixed bias for the IF stages. The drop across R_1 and R_2 is the bias value applied to the base of the IF stage. The drop across R_1 is combined with the pulsating detector signal output. Typically, the drop across R_1 is on the order of 0.5 to 1 V. The drop across R_1 and R_2 is between 1 and 2 V. The value of C_2 is quite large in relation to other bypass capacitors, and is typically 10 μF, or larger.

6-4. RF POWER AMPLIFIER AND MULTIPLIER DESIGN REQUIREMENTS

Figure 6-8 shows the working schematics of typical RF power amplifiers. The same basic circuits can be used as frequency multipliers. However, in a multiplier circuit, the output must be tuned to a multiple of the input. A multiplier may or may not provide amplification. Usually, most of the amplification is supplied by the final amplifier stage, which is not operated as a multiplier. That is, the input and output of the final stage are at the same frequency.

Sec. 6-4 RF Power Amplifier and Multiplier Design Requirements 185

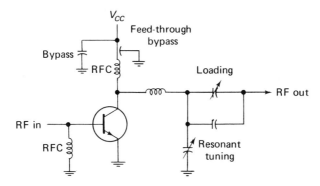

(a) Typical for Final Amplifier

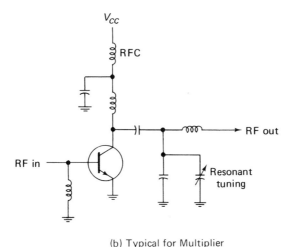

(b) Typical for Multiplier

Figure 6-8 Typical RF power amplifier and mutiplier circuits

A typical RF transmitter will have three stages: an oscillator to provide the basic signal frequency; an intermediate stage that provides amplification and/or frequency multiplication; and a final stage for power amplification. In some cases there are three stages of amplification following the oscillator stage.

6-4.1. Design Requirements

All of the general design considerations for other types of RF amplifiers apply to RF power amplifiers and multipliers. Of particular importance are interpreting datasheets, determining parameters at different frequencies, and temperature-related problems. In addition to the basic considerations, the following problems must be considered in RF power amplifier design.

Tuning and adjustment controls. The circuit of Fig. 6-8(a) has two tuning controls (variable capacitors) in the output network. The network of the Fig. 6-8(b) circuit has only one adjustment control. The circuit of Fig. 6-8(a) is typical for power amplifiers, where the output must be tuned to the resonant frequency by one control and adjusted for proper impedance match by the other control (often called the loading control). In practice, both controls affect tuning and loading (impedance matching). The circuit of Fig. 6-8(b) is typical for multipliers or intermediate amplifiers where the main concern is tuning to the resonant frequency.

Minimum capacitance. Also note that variable capacitors are connected in parallel with fixed capacitors in both networks. This parallel arrangement serves two purposes. First, it provides a minimum fixed capacitance in case the variable capacitor is adjusted to its minimum value. In some cases, if a minimum capacitance is not included in the network, a severe mismatch can occur when the variable capacitor is at its minimum, resulting in damage to the transistor. Another reason for the parallel capacitor is to reduce required capacitance rating (and thus the physical size) of the variable capacitor.

Tuning range. The maximum capacitance range of the tuning network is dependent upon the required tuning range of the circuit. A wide frequency range requires a wide capacitance range. In a practical circuit, use a capacitor with a midrange capacitance equal to the desired capacitance at the center frequency. For example, if 25 pF is required to produce resonance at the center operating frequency, use a variable capacitor with a range of 1 to 50 pF. If such a capacitor is not readily available, use a fixed capacitor of 15 pF in parallel with a 15-pF variable capacitor. This provides a capacitance range of 16 to 30 pF, with a midrange of about 23 pF.

Class of operation. Normally, amplifiers and multipliers using the circuits of Fig. 6-8 are operated as class B or C. The transistors remain cut off until a signal is applied, and are never conducting for more than 180° (half-cycle) of the 360° input signal cycle. In practice, the transistors conduct about 140° of the input cycle, either on the positive half or negative half, depending upon the transistor type (NPN or PNP). No bias, as such, is required for this class of operation.

Emitter connection. In RF power amplifiers, the emitter is connected *directly* to ground. In those transistors where the emitter is connected to the case (typical in many RF power transistors), the case can be mounted on a chassis that is connected to the ground side of the supply voltage. A direct connection between emitter and ground is of particular importance in high-frequency applications. If the emitter is connected to ground through a resistance (or even a long lead wire), an inductive or capacitive reactance can develop at high frequencies, resulting in undesired changes in the network. Another reason for direct connection between emitter and ground is to produce maximum gain. All other factors being equal, a decrease in emitter resistance (in relation to collector impedance) produces an increase in amplifier gain.

Power supply connections. The transistor base is connected to ground through an RF choke (RFC). This provides a d-c return for the base, as well as RF signal isolation between base and emitter or ground. The transistor collector is connected to the supply voltage through an RFC and (in some cases) through the coil portion of the resonant network. The RFC provides d-c return, but RF signal isolation, between collector and power supply. When the collector is connected to the power supply through the resonant network, the coil must be capable of handling the full collector voltage. For this reason, final amplifier networks are often chosen so that collector current does not pass through the coil (Fig. 6-8(a)). The circuit of Fig. 6-8(b) is used for applications where the current is low.

RFC ratings. The ratings for RFCs are sometimes confusing. Some manufacturers list a full set of characteristics: inductance, d-c resistance, a-c resistance, Q, current capability, and nominal frequency range; a-c resistance and Q are usually frequency-dependent. A nominal frequency range characteristic is helpful in design, but usually not critical.

All other factors being equal, the d-c resistance should be at a minimum for any circuit carrying a large amount of current. For example, a large d-c resistance at the collector of a final power amplifier can result in a large voltage drop between power supply and collector. Usually, the selection of a trial value for an RFC is based on a tradeoff between inductance and current capability. The minimum current capacity should be greater (by at least 10 per cent) than the maximum anticipated direct current. The inductance is dependent upon operating frequency. As a trial value, use an inductance that will produce a reactance between 1000 and 3000 Ω at the operating frequency.

Bypass capacitance. The power supply circuits must be bypassed. The feed-through bypass capacitors shown in Fig. 6-8 are used at higher frequencies where the RF circuits are physically shielded from the power supply and other circuits. The feed-through capacitor permits direct current to be applied through the shield, but prevents radio frequencies from passing outside the shield (radio frequencies are bypassed to the ground return). As a trial value, use a total bypass capacitance range of 0.001 to 0.1 μF.

From a practical standpoint, the best test for adequate bypass capacitance and RFC inductance is the absence of RF signals on the power supply side of the d-c voltage line. If RF signals are present on the power supply side, the bypass capacitance and/or RFC inductance is *not adequate*. (A possible exception to this is when the RF signals are being picked up due to inadequate shielding.) If the shielding is good and RF signals are present in the power supply, increase the bypass capacitance value. Also increase RFC inductance. Of course, circuit performance must be checked with each increase in capacitance or inductance value. For example, too much RFC inductance can reduce amplifier output and efficiency.

Efficiency. A class C RF amplifier has a typical efficiency of about 65 to 70 per cent. That is, the RF power output is 65 to 70 per cent of the d-c input power. To find the required d-c input power, divide the desired RF output by 0.65 or 0.7. For example, if the desired RF output is 50 W, the d-c input power is 50/0.7, or about 70 W. Ignore the slight voltage drop across the RFC and/or coil, and divide the input power by the power supply voltage to find the collector current. With a d-c input of 70 W and a 28 V supply, the collector current is 2.7 A.

Transistor characteristics. Transistors must be capable of handling the full power-supply voltage at their collectors, and the transistor current and/or power rating must be greater than the maximum required values. Likewise, the transistor must be capable of producing the necessary power output at the operating frequency.

The transistors must also provide the necessary *power gain at the operating frequency*. Likewise, the input power to an amplifier must match the desired output and gain. For example, assume that a 50 W RF amplifier is to be designed and that the available transistors have a power gain of 10. The final amplifier must have an input signal of at least 5 W with a d-c input of about 70 W.

Multiplier efficiency. Typically, the efficiency of a second harmonic amplifier or multiplier (with the output tuned to twice the input frequency) is about 40 per cent. The efficiencies of third, fourth, and fifth harmonic amplifiers are in the approximate range of 28, 21, and 18 per cent, respectively. Thus, if an IF amplifier is to be operated at the second harmonic and produce 5 W RF output, the d-c power input is approximately 12.5 W (5/0.40 = 12.5).

Another problem to be considered in frequency multiplication is that the power gain (as listed on the datasheet) may not remain the same as when amplifier input and output are at the same frequency. Some datasheets specify power gain at the basic frequency, and then derate the power gain for second harmonic operation. As a guideline, always use the *minimum power gain factor* when calculating power input and output values.

6-4.2. Resonant Network Analysis

In any RF amplifier, the tuning network must be resonant at the desired frequency. (Inductive and capacitive reactance must be equal at the selected frequency.) Also, the tuning network must match the transistor output impedance to the load.

Sec. 6-4 RF Power Amplifier and Multiplier Design Requirements

Generally, an antenna load impedance is about 50 Ω whereas the output impedance of a typical two-junction transistor at radio frequencies is only a few ohms. In the case where one amplifier feeds into another amplifier, the network must match the output impedance of one transistor to the input impedance of another transistor. Any mismatch can result in a loss of power between stages, or to the final load.

Transistor impedance (both input and output) has both resistive and reactive components, and therefore varies with frequency. To design a resonant network for the output of a transistor, it is necessary to know the output reactance (usually capacitive), the output resistance at the operating frequency, and the output power. Likewise, it is necessary to know the input resistance and reactance of a transistor at a given frequency and power when designing the resonant network of the stage feeding into the transistor.

Generally, the input resistance, the input capacitance, and the output capacitance of RF power transistors are shown by means of graphs similar to those of Fig. 6-9. The reactance can then be found using the corresponding frequency and capacitance. For example, the output capacitance shown on

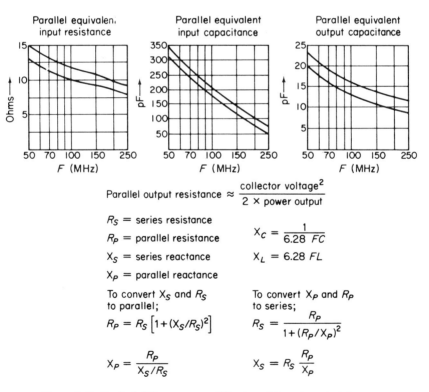

Figure 6-9 Typical RF power amplifier transistor characteristics

the graph of Fig. 6-9 is about 15 pF at 80 MHz. This produces a capacitive reactance of about 130 Ω at 80 MHz. The reactance and resistance can then be combined as necessary.

Input and output transistor impedances are generally listed on datasheets in *parallel form*. That is, the datasheets assume the resistance is in parallel with the capacitance. However, some tuning networks require that the impedance be calculated in series form. The equations necessary to convert between series and parallel forms are shown in Fig. 6-9.

The output resistance of RF power transistors is usually not shown on datasheets, but may be calculated using the equations of Fig. 6-9.

Typical resonant networks. Figures 6-10 through 6-14 show five typical resonant networks, together with the equations necessary to find component values. Note that these equations are best solved by computer-aided design techniques, since many are tedious and time consuming.

The networks can be used as amplifiers and/or multipliers. Note that the

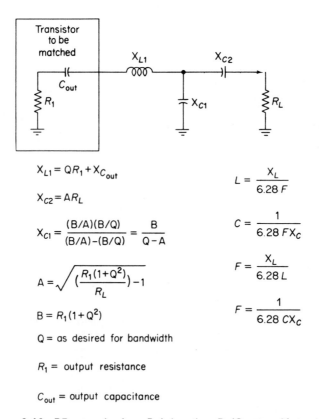

$X_{L1} = QR_1 + X_{C_{out}}$

$X_{C2} = AR_L$

$X_{C1} = \dfrac{(B/A)(B/Q)}{(B/A)-(B/Q)} = \dfrac{B}{Q-A}$

$A = \sqrt{(\dfrac{R_1(1+Q^2)}{R_L})-1}$

$B = R_1(1+Q^2)$

$Q =$ as desired for bandwidth

$R_1 =$ output resistance

$C_{out} =$ output capacitance

$L = \dfrac{X_L}{6.28\,F}$

$C = \dfrac{1}{6.28\,FX_C}$

$F = \dfrac{X_L}{6.28\,L}$

$F = \dfrac{1}{6.28\,CX_C}$

Figure 6-10 RF network where R_1 is less than R_L (Courtesy Motorola)

network of Fig. 6-10 is similar to that of Fig. 6-8(a), whereas Fig. 6-11 is similar to Fig. 6-8(b).

Impedance matching. The resistor and capacitor shown in the box labeled "Transistor to be matched" represent the *complex output impedance* of a transistor. When the network is to be used with a final amplifier, the resistor labeled R_L is the antenna impedance, or other load. When the network is used with an intermediate amplifier, R_L represents the input impedance of the following transistor. It is therefore necessary to calculate the input impedance of the transistors being fed by the network, using the data and equations of Fig. 6-9.

The complex impedances are represented in series form in some cases and parallel form in others, depending on which form is the most convenient for network calculation. The resultant impedance of the network, when terminated with a given load, must be equal to the conjugate of the impedances in the box. For example, assume that the transistor has a series output

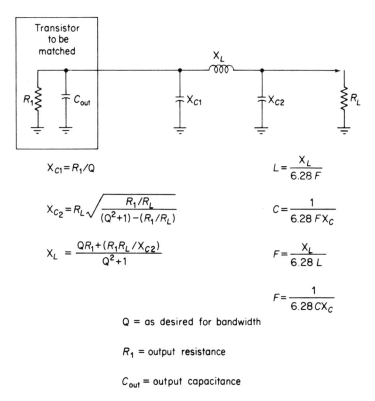

$$X_{C1} = R_1/Q$$

$$X_{C2} = R_L \sqrt{\frac{R_1/R_L}{(Q^2+1)-(R_1/R_L)}}$$

$$X_L = \frac{QR_1 + (R_1 R_L / X_{C2})}{Q^2 + 1}$$

$$L = \frac{X_L}{6.28 F}$$

$$C = \frac{1}{6.28 F X_C}$$

$$F = \frac{X_L}{6.28 L}$$

$$F = \frac{1}{6.28 C X_C}$$

Q = as desired for bandwidth

R_1 = output resistance

C_{out} = output capacitance

Figure 6-11 RF network where R_1 is approximately equal to R_L (Courtesy Motorola)

impedance of $8.77 - j9.3$. That is, the resistance (real part of impedance) is 8.77 Ω, whereas the capacitive reactance (imaginary part of impedance) is 9.3 Ω. If this was expressed as a y-parameter (output admittance, or y_{22}), it would be g (114 mmhos) $- jb$ (107 mmhos). However, the networks of Figs. 6-10 through 6-14 are best solved using impedances (resistance and reactance).

For maximum power transfer from the transistor to the load, the load impedance must be the conjugate of the output impedance, or $8.77 + j9.3$. If the amplifier is designed to operate into a 50-Ω external load, the network must transfer the $50 + j0$ external load to the $8.77 + j9.3$ transistor load.

In addition to performing this transformation, the network provides harmonic rejection (unless a harmonic is needed in a multiplier stage), low loss, and provisions for adjustment of both loading and tuning.

Network characteristics. Each network has advantages and limitations. The network of Fig. 6-10 applies to most RF power amplifiers and is especially useful when the series real part of the transistor output impedance, or R_1, is less than 50 Ω. With a typical 50-Ω load, the required reactance of C_1 rises to an impractical value when R_1 is close to 50 Ω.

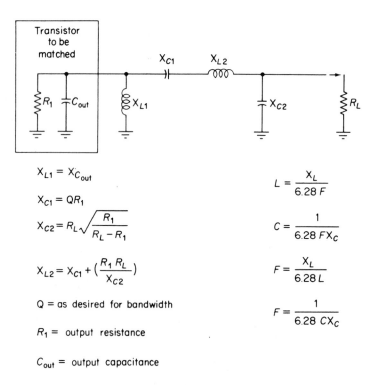

Figure 6-12 RF network where R_1 is very small in relation to R_L (Courtesy Motorola)

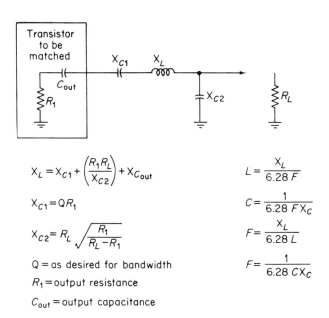

$$X_L = X_{C1} + \left(\frac{R_1 R_L}{X_{C2}}\right) + X_{Cout}$$

$$X_{C1} = QR_1$$

$$X_{C2} = R_L \sqrt{\frac{R_1}{R_L - R_1}}$$

Q = as desired for bandwidth
R_1 = output resistance
C_{out} = output capacitance

$$L = \frac{X_L}{6.28\,F}$$

$$C = \frac{1}{6.28\,F X_C}$$

$$F = \frac{X_L}{6.28\,L}$$

$$F = \frac{1}{6.28\,C X_C}$$

Figure 6-13 RF network where R_1 is very small in relation to R_L (alternative network) (Courtesy Motorola)

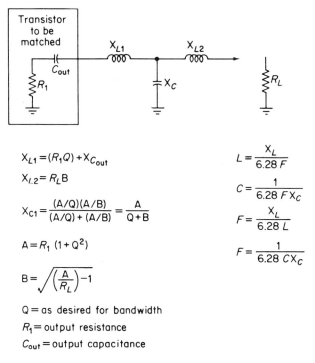

$$X_{L1} = (R_1 Q) + X_{Cout}$$

$$X_{L2} = R_L B$$

$$X_{C1} = \frac{(A/Q)(A/B)}{(A/Q) + (A/B)} = \frac{A}{Q+B}$$

$$A = R_1 (1 + Q^2)$$

$$B = \sqrt{\left(\frac{A}{R_L}\right) - 1}$$

Q = as desired for bandwidth
R_1 = output resistance
C_{out} = output capacitance

$$L = \frac{X_L}{6.28\,F}$$

$$C = \frac{1}{6.28\,F X_C}$$

$$F = \frac{X_L}{6.28\,L}$$

$$F = \frac{1}{6.28\,C X_C}$$

Figure 6-14 RF network where R_1 is very small, or very large, in relation to R_L (Courtesy Motorola)

The network of Fig. 6-11 (often known as a Pi network) is best suited where the parallel resistance R_1 is high (near the value of R_L, typically 50 Ω). If the network of Fig. 6-11 is used with a low value of R_1 resistance, the inductance L_1 must be very small, whereas C_1 and C_2 become very large (beyond practical limits).

The networks of Figs. 6-12 and 6-13 produce practical values of C and L, especially where R_1 is very low. The main limitation for the networks of Figs. 6-12 and 6-13 is that R_1 must be substantially lower than R_L. These networks, or variations thereof, are often used with intermediate stages where a low output impedance of one transistor must be matched to the input impedance of another transistor.

The network of Fig. 6-14 (often known as a Tee network) is best suited where R_1 is much less or much greater than R_L. That is, the Fig. 6-14 network is well suited to drastic mismatch situations (such as matching a transistor with an output impedance of 10 Ω to a 300-Ω antenna).

6-5. RF AMPLIFIER DESIGN WITH DATASHEET GRAPHS

High-frequency characteristics are especially important in the design of RF networks. Unfortunately, the high-frequency information provided on many datasheets in tabular form is not adequate for simplified design. To properly match impedances, both the resistive and reactive components must be considered. The reactive component (either inductive or capacitive) changes with frequency. In practical amplifier work, it is necessary to know the reactance values over a wide range of frequencies, not at some specific frequency (unless you happen to be designing for that frequency only).

The best way to show how resistance and reactance vary in relation to frequency for a particular transistor is by means of graphs or curves. Fortunately, manufacturers who are trying to sell their transistors for high-frequency power-amplifier use generally provide a set of curves showing the characteristics over the anticipated frequency range.

The following paragraphs describe the basic design steps for an RF power amplifier, using typical datasheet curves and the equations of Figs. 6-9 through 6-14.

The amplifier shown in Fig. 6-15 delivers 25 W output at 175 MHz, with a gain of about 21 dB and 47 per cent overall efficiency.

6-5.1. Circuit Analysis

A summary of the amplifier's performance is given in the table in Fig. 6-15. The circuit operates from a negative-grounded 12.5 V supply

Sec. 6-5 RF Amplifier Design with Datasheet Graphs 195

Circuit Performance	
RF power output	= 25 W
RF power input	= 190 mW
Supply voltage	= 12.5 V d-c
Total current	= 4.3 A
Overall efficiency	= 46.5%
All harmonics and spurious outputs	40 dB below fundamental output
R_G	= 50 Ω
R_L	= 50 Ω

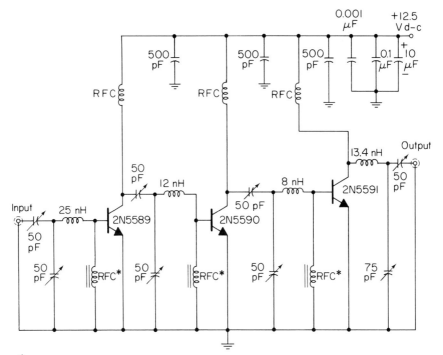

*All base chokes are Ferroxcube ferrite type VK-200 19/4B

Figure 6-15 25-W 175-MHz RF amplifier circuit (Courtesy Motorola)

and draws 4.3 A for a 25-W output. All stages are class C amplifiers with zero no-signal bias. Conventional single-tuned networks are used throughout.

The amplifier uses the Motorola 2N5589 through 2N5591 NPN silicon VHF power transistors designed specifically for operation directly from a 12-V vehicular electrical system. The gain and RF power output capabilities of these transistors are shown in Figs. 6-16 through 6-18. All three transistors are supplied in a low-inductance radial-lead stud package for improved RF performance and ease of mounting.

Basic design procedure. The design approach used for this amplifier (Fig. 6-15) involves the use of the large-signal transistor impedances as a basis for synthesis of the matching networks. The impedances referred to are the terminal impedances of the transistor when operated in an optimized power amplifier at rated power output and supply voltage and should not be confused with small-signal two-port parameters (y-parameters, for example), which are normally measured at low signal levels with short, open, or 50-Ω terminations.

The large-signal impedances of the 2N5589-91 transistors have been measured and are shown in Figs. 6-19 through 6-21, and on the Motorola device datasheets. This data was measured at the power outputs and supply voltages indicated with the transistors operating in the common-emitter configuration with the emitter at d-c ground.

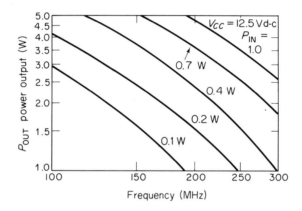

Figure 6-16 Power output versus frequency for 2N5589 (Courtesy Motorola)

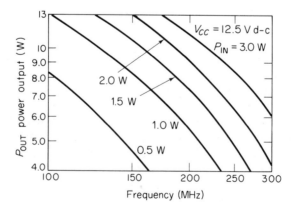

Figure 6-17 Power output versus frequency for 2N5590 (Courtesy Motorola)

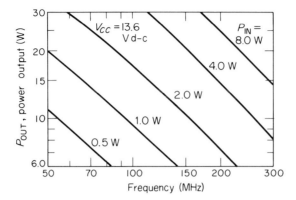

Figure 6-18 Power output versus frequency for 2N5591 (Courtesy Motorola)

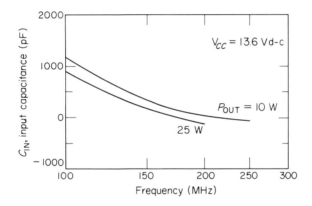

Figure 6-19 Parallel equivalent input capacitance versus frequency for 2N5591 (Courtesy Motorola)

Transistor impedances. Both resistive (R_{IN}) and reactive (C_{IN}) components of the parallel input impedance are given in the curves in Figs. 6-19 and 6-20. For output impedance, only the parallel output capacitance (C_{OUT}) is given (in Fig. 6-21). The resistive portion of the collector load impedance (R'_L) may be computed for any given power output and supply voltage (V_{CC}) by assuming a peak-to-peak collector voltage swing of twice the V_{CC}. These conditions yield the following expression for R'_L (See Fig. 6-9).

$$R'_L = \frac{(V_{CC})^2}{2P}$$

where P = RF power output.

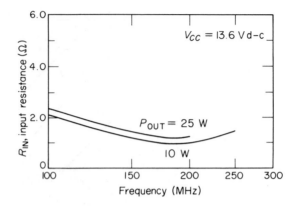

Figure 6-20 Parallel equivalent input resistance versus frequency for 2N5591 (Courtesy Motorola)

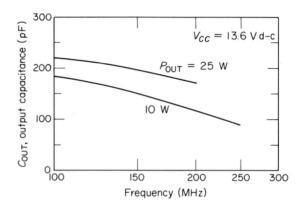

Figure 6-21 Parallel equivalent output capacitance versus frequency for 2N5591 (Courtesy Motorola)

Collector load. The actual optimum collector load resistance may vary somewhat from the value computed for R'_L. However, designing the output network for a conjugate match to R'_L (computed from the equation) in parallel with C_{OUT} has yielded excellent results with a large variety of transistor types, power-output levels, and frequencies.

Output network. The output network configuration shown in Fig. 6-10 was selected because it performs the required impedance matching with convenient component sizes. Also, the network tunes smoothly and provides good harmonic attenuation.

Calculating conjugate values. The output capacitance of the 2N5591 when operating at 25-W output is 185 pF, as shown in Fig. 6-21. Note that the

Sec. 6.5.   RF Amplifier Design with Datasheet Graphs   199

impedance data is given for 13.6 V operation. However, operating the 2N5591 at 12.5 V instead of 13.6 V does not change the transistor's impedance characteristics significantly, and the data given may be used for 12.5-V amplifier designs.

R'_L may be computed from the Fig. 6-9 equation as follows

$$R'_L = \frac{(12.5)^2}{2 \times 25} = 3.13 \, \Omega$$

Therefore, the output network must be designed to transform the 50-Ω external load to the conjugate of 3.13-Ω resistance in *parallel* with a capacitance of 185 pF.

In the design of the output network with the configuration shown in Fig. 6-10, the *series* form of collector load impedance is more convenient. The collector load impedance of 3.13 Ω in parallel with 4.9-Ω inductive reactance (the conjugate of 3.13-Ω resistance in parallel with 185-pF capacitance at 175 MHz) may be converted to series form using the equations (See Fig. 6-9)

$$R_S = \frac{R_P}{1 + \left(\frac{R_P}{X_P}\right)^2} = \frac{3.13}{1 + \left(\frac{3.13}{4.9}\right)^2} = 2.22 \, \Omega$$

$$X_S = R_S \frac{R_P}{X_P} = \frac{2.22 \times 3.13}{4.9} = 1.42 \, \Omega$$

Therefore, the required collector load impedance is $2.22 + j1.42 \, \Omega$.

Network Q. For purposes of network design, the loaded $Q(Q_L)$ of the output network may be defined as X_{L1}/R_S, where R_S is the series equivalent collector load resistance.

A Q_L of 6 was selected for the 2N5591 output network. This value provides a good combination of harmonic attenuation and low insertion loss.

X_{L1} is computed as

$$X_L = Q_L \times R_S = 6 \times 2.22 = 13.2 \, \Omega$$

Series inductor. An additional series inductor is required between the collector and L_1 to tune out the series equivalent output capacitive reactance of the 2N5591, which was computed to be 1.42 Ω.

In practice, this additional inductor may be combined with L_1 to form a single series inductor with a *total reactance* of $13.3 + 1.42$, or $14.7 \, \Omega$. This requires an inductance of 13.4 nH at 175 MHz.

Network capacitance. X_{C1} and X_{C2} may then be computed using the equations of Fig. 6-10. Solving the equations and converting the reactances to capacitances yields: $C_1 = 56$ pF, $C_2 = 23$ pF.

The values for remaining networks in the amplifier of Fig. 6-15 can be found using similar procedures.

6-6. RF VOLTAGE AMPLIFIER DESIGN

This section is devoted to design of RF voltage amplifiers. It is possible to design such amplifiers, both two-junction and FET, using two-port systems and admittance parameters, as described in Sec. 2-2. Although the examples given in this section are for FETs, the same basic principles apply to two-junction transistors.

6-6.1. Design Example of a 200-MHz Neutralized MOSFET RF Amplifier

Assume that the circuit of Fig. 6-22 is to be operated at 200 MHz. The source and load impedances are both 50 Ω. The characteristics of the MOSFET (taken from the datasheet) are as follows (all admittance values in mmhos)

$$y_{11} = 0.45 + j7.2$$
$$y_{22} = 0.28 + j1.75$$
$$y_{21} = 7.0 - j1.9$$
$$y_{12} = 0 - j0.16$$

Note that the real part of y_{12} is considered as zero. c_{rss} is 0.2 pF. Tuning is to be accomplished with standard 1 to 9-pF variable capacitors. It is assumed that the bias networks have been designed as described in Chapter 5. No special considerations need be given to bandwidth or selectivity in this example.

When the y-parameters are plugged into the Linvill equation, the Linvill stability factor C is over 2.

$$C = \frac{|(7.0 - j1.9)(0 - j0.16)|}{2(0.45)(0.28) - R_e(7.0 - j1.9)(-0.16)} \approx 2.08$$

Thus, the MOSFET is not unconditionally stable and neutralization may be required. It is possible that a mismatch between the MOSFET input-output impedances and the 50-Ω source/load impedances would produce sufficient stability. However, to be sure, neutralization is used.

Since neutralization is used, the amplifier can be conjugately matched for *maximum gain*. With the real part of y_{12} assumed to be zero ($g_{12} = 0$) and a conjugate match, the maximum available gain expression can be used to find an MAG of about 20 dB.

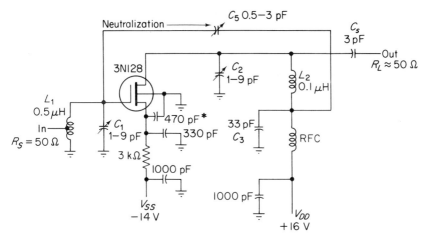

Figure 6-22 200-MHz common-source amplifier circuit with neutralization (Courtesy RCA)

$L_1 = 4\frac{1}{2}$ turns of No. 20 wire, $\frac{3}{16}$ in. diam, $\frac{1}{2}$ in. long, tapped at 1 turn

$L_2 = 3\frac{1}{2}$ turns of No. 20 wire, $\frac{3}{8}$ in. diam, $\frac{1}{2}$ in. long

*Leadless disc capacitor

$$\text{MAG} = \frac{7.0 - j1.9^2}{4(0.45)(0.28)} = 104 = 20.2 \text{ dB}$$

The source and load impedances must be matched to the MOSFET input and output impedances, respectively, to obtain the maximum gain. The input is matched by means of transformation (coupling autotransformer turns ratio). The output is matched by means of coupling capacitor reactance.

The source input is 50 Ω. For a conjugately matched input, the source conductance G_s and the real part of the MOSFET input admittance must *appear* to be equal. Source conductance is the reciprocal of 50 Ω, or 20 mmhos. The real part of the MOSFET input admittance $R_e(y_{11})$ is 0.45 mmho. The transformation ratio is approximately 44 (20/0.45), with a turns ratio of about 6.6 ($\sqrt{44}$). Experimentally, a turns ratio of about 4 was found to be approximately the best value. The difference partially results from the fact that the parallel resistance of the tank coil was not considered in the calculation.

Assuming that the input tuning capacitor is set to a value between 1 and 2 pF, the required input inductance is about 0.5 μH.

$$L \text{ (in } \mu\text{H)} = \frac{2.54 \times 10^4}{\text{freq (in MHz)}^2 \times \text{capacitance (in pF)}}$$

The load output is also 50 Ω. Because the d-c drain voltage must be blocked from the load, a series coupling capacitor C_S must be used. Capacitor C_S also functions to match the MOSFET output to the load. The load conductance G_L is 20 mmhos, and the real part of the MOSFET output admittance, $R_e(y_{22})$, is 0.28 mmho. The reactance for C_S that will produce a conjugate match is found by

$$XC_S = R_S \sqrt{\frac{R_P}{R_S} - 1}$$

where R_P is the reciprocal of the parallel MOSFET output admittance or $1/R_e(y_{22})$, R_S is the reciprocal of the load conductance, $1/G_S$ (or the load impedance), and R_P is much greater than R_S.

Using these values, the reactance of C_S is

$$X_{CS} = 50 \sqrt{\frac{3600}{50} - 1} \approx 420 \text{ } \Omega$$

At 200 MHz, the value for C_S is found by

$$C_S = \frac{1}{6.28 \times (200 \times 10^6) \times 420} \approx 1.9 \text{ pF}$$

Experimentally, a 3-pF capacitor was found to perform satisfactorily.

The capacitance of C_S is in parallel with the output capacitance of the MOSFET. As a guideline, the MOSFET output capacitance can be found when the imaginary part of the output admittance jb_{22} is put into the following equation

$$\text{Parallel capacitance} = \frac{1}{6.28 \times (200 \times 10^6) \times \left(\frac{1}{1.75} \text{ mmhos}\right)} \approx 1.4 \text{ pF}$$

Note that the output admittance of the MOSFET does not truly equal y_{22}, except when the input is shorted, and y_{12} is truly zero. Thus, the jb_{22} figure is not necessarily accurate. However, the jb_{22} figure can be used to find a trial value for design purposes in this example. If bandwidth and/or selectivity are of concern, a more exact figure for y_{22} is required. This figure can be found using the output admittance equations of Sec. 2-2.3.

In this example, a 1.4-pF output capacitance is considered to be in parallel with the 3-pF capacitance of C_S, resulting in a total capacitance of 4.4 pF. To simplify calculations, the variable tuning capacitor is considered as set to 2 pF. This produces a total parallel output capacitance of 6.4 pF.

The required output inductance is about 0.1 μH, found from the equation

$$L \text{ (in } \mu\text{H)} = \frac{2.54 \times 10^4}{(200)^2 \times 6.4} \approx 0.1 \text{ } \mu\text{H}$$

Sec. 6-6   RF Voltage Amplifier Design   203

An approximate value for C_N is found when c_{rss} is multiplied by the ratio of C_3 to C_2. The value of C_2 is the total parallel output capacitance of 6.4 pF. The value of C_3 must be of at least four times that of C_2. In this example, a standard value of 33 pF is chosen ($5 \times 6.4 = 32$). With a c_{rss} of 0.2, the value of C_N is

$$C_N \approx 0.2 \times \left(\frac{33}{6.4}\right) \approx 1 \text{ pF}$$

In practice, a standard variable capacitor ($0.5 - 3$ pF) is used for C_N. This makes it possible to adjust the feedback (neutralization) for variations in c_{rss}. Typically, the c_{rss} for the 3N128 varies between 0.15 and 0.35 pF.

6-6.2. Design Example of 100-MHz Front-end MOSFET RF Amplifier for FM Tuner.

Figure 6-23 shows the circuit of a typical MOSFET RF amplifier used as the front-end of an FM tuner. Because of their low harmonic output, MOSFETs are well suited as front-end RF amplifiers in tuners.

Assume that the circuit of Fig. 6-23 is to be operated at 100 MHz, and

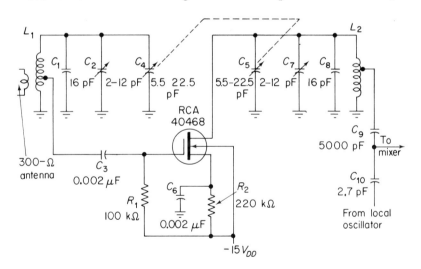

$L_1 = \#18$ bare copper wire, 4 turns, $\frac{1}{4}$ in. ID, $\frac{7}{17}$ in. winding length, Q at 100 MHz = 120. Tunes with 34 pF at 100 MHz. Antenna link approx. 1 turn from ground end. Gate tap approx. $1\frac{1}{2}$ turns from ground end.

L_2 = Same as L_1 except mixer tap at approximately $\frac{3}{4}$ turn.

Figure 6-23 Typical FM receiver RF amplifier using 40468 MOSFET (Courtesy RCA)

that the bias networks have been designed as described in Chapter 5. The values of circuit components obtained by means of the following design are given in the parts list of Fig. 6-23. The problem here is to find the best ratios for matching input and output circuits to the amplifier.

The following parameters are important in the design of the RF amplifier stage

MOSFET (40468) parameters (at $V_{DD} = 15$ V, $I_D = 5$ mA):
    input resistance R_{IN} 4500 Ω
    output resistance R_{OUT} 4200 Ω
    forward transadmittance y_{fs} 7500 μmhos
    feedback capacitance c_{rss} (max)   0.2 pF

Mixer stage parameters:
    input resistance R_{IN} 550 Ω
    input stability $IS_{(\text{mix})}$ 4

Coil data:
    mounted unloaded Q 120
    tuning capacitance C_T at 100 MHz   34 pF
    antenna impedance   300 Ω

Figure 6-24 shows the a-c equivalent circuit for the RF stage. At resonance, this circuit reduces to the form shown in Fig. 6-25, where all impedances are referred to the gate and drain terminals of the MOSFET.

Using the maximum available gain (MAG) equation of Sec. 2-2.3 and the values for the MOSFET,

$$\text{MAG} = \frac{(7500 \times 10^{-6})^2 \times 4500 \times 4200}{4} \approx 266 \approx 24.2 \text{ dB}$$

Using the maximum usable gain (MUG) equation of Sec. 2-2.3 and the

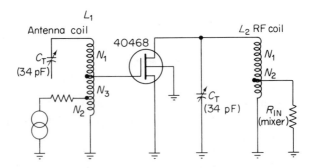

Figure 6-24   Equivalent circuit for RF amplifier stage (Courtesy RCA)

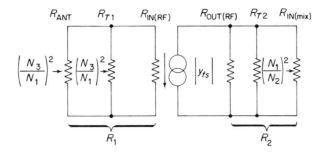

Figure 6-25 Equivalent input R_1 and output R_2 circuit of the RF stage at resonance (Courtesy RCA)

values for the MOSFET,

$$\text{MUG} = \frac{0.4 \times (7500 \times 10^{-6})}{6.28(10^8) \times (0.2 \times 10^{-12})} \approx 23.5 \approx 13.7 \text{ dB}$$

The total mismatch loss is called the *stability factor S*, and is equal to the difference (in decibels) between MAG and MUG, as follows

$$S = \text{MAG} - \text{MUG} = 24.2 - 13.7 = 10.5 \text{ dB (or 11.3 times)}$$

In the design scheme of this example, the S value is divided between the input and output circuits by means of an *input stability factor IS* and an *output stability factor OS*, as follows

$$IS = \frac{R_{IN}}{2R_1}, \qquad OS = \frac{R_{OUT}}{2R_2}$$

where R_1 and R_2 are the total parallel impedances of the input terminal (gate) and output terminal (drain), respectively. (See Fig. 6-25.)

These stability terms are related to the stability factor S as follows

$$S = IS \times OS, \qquad IS = \frac{S}{OS}, \qquad OS = \frac{S}{IS}$$

The division between IS and OS is made by some arbitrary choices. In this example, IS is maximized so that the signal level at the gate will be minimized. This choice requires matching (or nearly matching) R_{OUT} to its load. Therefore, the entire RF coil (L_2) is used as the output load.

To get an impedance match between R_{OUT} and the load, let $OS = 1$, or unity. Then $IS = S/OS$, or $IS = 11.3/1 = 11.3$.

With $OS = 1$, the value of R_2 is found by $OS = R_{OUT}/2R_2$, $2R_2 = R_{OUT}/S$

$$R_2 = \frac{\left(\frac{R_{OUT}}{OS}\right)}{2} = \frac{\left(\frac{4200}{1}\right)}{2} = 2100 \; \Omega$$

With $R_2 = 2100$, the ratio of N_1/N_2 is found by

$$\frac{N_1}{N_2} = \sqrt{\frac{IS_{(mix)} \; 2R_2}{R_{IN \; (mix)}}} = \frac{4 \times 4200}{550} \approx 5.5$$

In practice, on a four-turn coil L_2, this ratio is accomplished by making the tap at approximately 3/4 turn from the ground end.

To match the 300-Ω antenna with the input circuit, the value of R_1 should be one-half the *reflected* antenna impedance.

With $IS = 11.3$, the value of R_1 is found by

$$IS = \frac{R_{IN}}{2R_1}, \quad 2R_1 = \frac{R_{IN}}{IS}, \quad R_1 = \frac{\left(\frac{R_{IN}}{IS}\right)}{2} = \frac{\frac{4500}{11.3}}{2} \approx 200 \; \Omega$$

This value for R_1 is so much lower than R_{IN} (4500), it can be seen that the MOSFET does not load the antenna coil excessively. Also, the reflected antenna impedance is about 400 Ω, if R_1 is one-half the reflected value.

The ratio N_1/N_3 must match the reflected antenna impedance of about 400 Ω to the tuned impedance of the input coil RT_1. The ratio N_1/N_2 must match the actual antenna impedance of 300 Ω to RT_1. The value of RT_1 is dependent upon the unloaded Q (given as 120), the tuning capacitance (given as 34 pF), and the frequency (100 MHz). The relationship is

$$RT_1 = \frac{Q}{6.28F \times C_T} = \frac{120}{6.28(10^8) \times (34 \times 10^{-12})} \approx 5600 \; \Omega$$

The N_1/N_3 transformation ratio is approximately 400/5600, or 14. This requires a turns ratio of about 3.7. The N_1/N_2 transformation ratio is approximately 300/5600, or 18.6. This requires a turns ratio of about 4.3.

In practice, on a four-turn coil L_1, these ratios are accomplished by making the taps at approximately one turn from the ground end (for antenna tap N_2) and 1.5 turns from the ground end (for gate tap N_3).

7 TRANSISTOR SWITCHES

Two-junction transistors and FETs are used in many applications as switches. The transistor choppers used in low-drift amplifiers are classic examples. Two-junction transistors are also used as switches in inverter or converter solid-state power supplies. In this chapter we shall discuss the general switching characteristics of transistors, but will concentrate on transistors used as choppers, inverters, and analog switches.

7-1. BASIC CHOPPER CIRCUITS

Chopper circuits are basically of either the *shunt type* or the *series type*, as shown in Fig. 7-1, or a combination of the two.

The shunt chopper circuit, shown in Fig. 7-1(a), operates as follows: When the switch S is opened, a voltage that is directly proportional to the input signal appears across the load. When the switch is closed, all of the input signal is shorted to ground. If the switch is opened and closed periodically, the voltage across the load appears as a square wave that has an amplitude directly proportional to the input signal. This square wave may be highly amplified by a relatively drift-free, stable-gain a-c amplifier. This procedure is generally used in low-level d-c amplifiers, where a small d-c input is chopped; the resulting a-c signal is amplified, and the output of the a-c amplifier is rectified to produce a d-c output directly proportional to the input.

The series chopper circuit, shown in Fig. 7-1(b), can also be used to chop d-c signals. This type of circuit is particularly useful in telemetry or other systems where a signal source such as a transducer is to be connected periodically to a load.

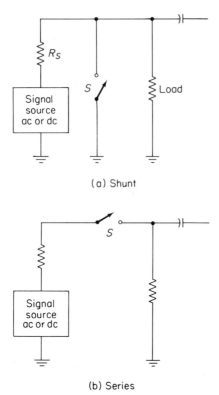

Figure 7-1 Basic chopper circuit

An ideal chopper is simply an on-off switch that has certain desirable characteristics. Figure 7-2 lists some of these characteristics and shows the relative merits of relays, two-junction transistors, and FETs in each area.

Early choppers were the mechanical types and used relays. These relay-type chopper circuits are characterized by near ideal contact performance; that is, extremely high impedance when open, and near-zero impedance when closed. However, these advantages are compromised by several disadvantages. Relays have short lifetimes, slow switching speeds, contact bounce, and generally, large size. Any electrical overstress might also result in an increase in contact resistance.

The disadvantages of relays in choppers are eliminated by chopper transistors, but not without introducing new disadvantages. For example, a transistor cannot match the near-ideal switching characteristics of a relay. A two-junction transistor has the additional disadvantage of offset voltage and offset current (which can be minimized through proper circuit design). In the case of FETs, particularly MOSFETs, there is a tendency for the chopper drive signal to feed through onto the signal being chopped. Again, this problem can be minimized by proper circuit design. All of these factors are discussed in the following paragraphs.

Ideal Chopper Characteristics	Two-Junction	FET	Electromechanical Relay
Infinite life	good	good	poor
Infinite frequency response	good	good	poor
Infinite OFF resistance	fair	good	good
Zero ON resistance	fair	poor	good
Zero driving-power consumption	fair	good	fair
Zero offset voltage	poor	good	good
Zero feedthrough between the driving signal and signal being chopped	fair	fair	good
Small size	good	good	poor

Figure 7-2 Comparison of available chopper devices with an ideal

7-2. TWO-JUNCTION TRANSISTOR CHOPPERS

If a common-emitter transistor switch (Fig. 7-3) is in the ON state, but the collector current is near zero, the collector-emitter voltage will not be zero. It is this *offset voltage* that most seriously limits the performance of two-junction transistors in chopper circuits. The circuit that follows the chopper will not be able to distinguish between the offset voltage and signal levels of comparable magnitude.

In the OFF state, the transistor will not be an ideal open circuit, but will appear as a current source. This *offset current* can induce a voltage in the load, again masking signal levels of comparable level.

The design problem is one of eliminating, or more realistically, reducing, the offset voltage and offset current. The most popular technique is to use

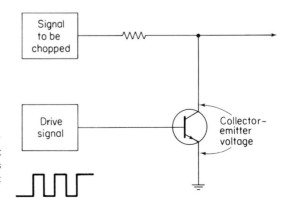

Figure 7-3 Common-emitter transistor switch showing effect of collector-emitter voltage (as an offset voltage and offset current source)

common-collector, or *inverted*, operation rather than common-emitter. (By inverted connection it is meant that the control signal or drive signal is applied between base and collector, instead of between base and emitter.)

The common-collector versions of shunt and series chopper circuits are shown in Figs. 7-3 and 7-4, respectively. The pulse transformer shown in Fig. 7-4 is required to avoid biasing the output signal with the chopping signal. The electrostatic shield reduces capacitive coupling of switching transients through the transformer to the load.

In the ON state of the series chopper (Fig. 7-4(b)), the voltage appearing across the transistor is not zero. Instead, the voltage is the equivalent of V_{EC} and $i_b r'_c$. This voltage represents a minimum limit for the size of the signal (e_s) that can be supplied to the load.

In the OFF state (Fig. 7-4(c)), the transistor appears as a current source, I_E, shunted by the leakage resistance across the emitter junction. It is the flow of I_E through the load, R_L, that gives an offset during the OFF state.

In the transient states between ON and OFF, the transistor base is coupled to the emitter and collector by junction capacitance. For example, assume

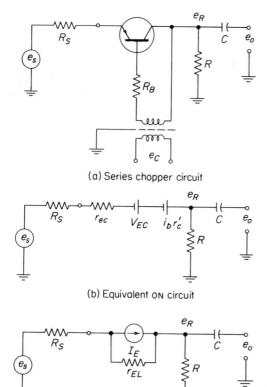

Figure 7-4 Basic two-junction transistor series chopper circuit (Courtesy Motorola)

that the OFF bias from collector to base is 6 V (the chopping signal is 12 V peak-to-peak at the transformer). When the chopping signal suddenly swings 12 V to the -6 V ON value, the junction capacitance tries to pass this 12-V transient. The actual value of the transient that appears at the collector will be a function of current limiting in the circuit. However, this transient can be significant. A similar situation appears during the turn-off switching time.

To summarize, the disadvantages of either the shunt or series choppers include (1) offset voltage, (2) offset current, and (3) transient feedthrough.

7-2.1. Practical Two-Junction Chopper Circuits

While the series chopper of Fig. 7-4 might be satisfactory in some noncritical applications, the demands of more sophisticated systems require reduction of offset errors. The series-pair chopper circuit of Fig. 7-5 is designed for this reduction.

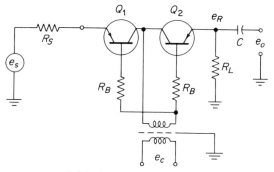

(a) Series pair chopper circuit

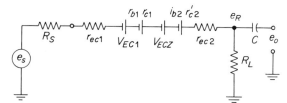

(b) Equivalent ON circuit

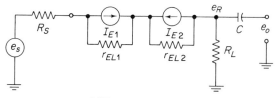

(c) Equivalent OFF circuit

Figure 7-5 Basic two-junction series-pair chopper circuit (Courtesy Motorola)

As the equivalent ON circuit (Fig. 7-5(b)) shows, each transistor contributes an offset voltage, but the voltages are of opposite polarity and tend to cancel. If a very tightly matched pair of transistors is used, the effective offset voltage will be almost zero. That is, for an input voltage, e_s, of 0, the output, e_o, will be 0.

The OFF equivalent in Fig. 7-5(c) shows that the offset currents are opposing. Thus, the net offset current through the load will be the difference between I_{E1} and I_{E2}. Again, a tight match will yield a zero difference, or a zero offset current error. While it might be possible to find closely matched pairs at 25°C, the match may degrade with age and temperature. Thus, a drift error will appear. The drift problem is discussed in later paragraphs of this section.

Figure 7-6 shows an improved version of the series-pair chopper. The diode D_1 prevents the collector-base junctions from becoming reverse-biased. This will eliminate the OFF leakage currents. The diode itself will have some leakage, but this is passed by R_1, rather than the transistors. The small capacitor C_1 forms a capacitive divider with the transistor junction capacitance. This helps suppress feedthrough transients.

The improvements in Fig. 7-6 are not without disadvantages. The zero off-bias mode and capacitive divider circuits will both tend to increase switching times. A tradeoff between speed, leakage, and feedthrough must be made.

Another circuit designed to reduce the effects of offsets is the series-shunt chopper shown in Fig. 7-7. The chopping signal is so coupled that while the series device is ON, the shunt device is OFF and vice-versa. These two states are shown in the equivalent circuits of Figs. 7-7(b) and 7-7(c).

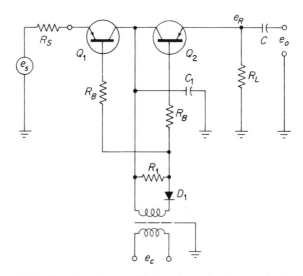

Figure 7-6 Two-junction modified series-pair chopper circuit (Courtesy Motorola)

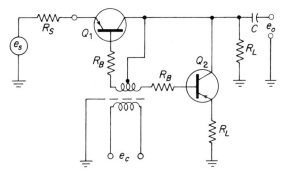

(a) Series-shunt chopper circuit

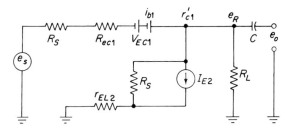

(b) Q_1 ON, Q_2 OFF equivalent circuit

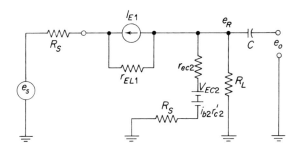

(c) Q_1 OFF, Q_2 ON equivalent circuit

Figure 7-7 Two-junction series-shunt chopper circuit (Courtesy Motorola)

If the devices are tightly matched, the voltage across R_L (exclusive of signal voltage) will be constant. Therefore, the capacitive coupling of this voltage will result in no offset appearing at e_o. Again, a loss of match due to age or temperature drift can result in some offset appearing at e_o. A control of drift offset and feedthrough of transients can be achieved by lowering R_L. However, this is not always practical.

In the circuits discussed thus far, it has generally been assumed that the impedance of the signal source R_S is low. Under this assumption, any offset

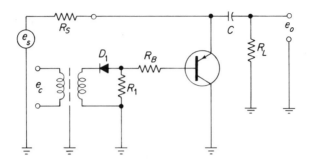

Figure 7-8 Two-junction shunt chopper circuit (Courtesy Motorola)

voltage due to leakage current flowing through the source impedance is also low. There are instances when the source impedance will be high. Under this condition, the leakage current-source impedance product may become intolerable.

By using a shunt chopper with zero off-bias, the $I_E R_S$ product can be eliminated. Figure 7-8 shows such a circuit. During the OFF time, the transistor looks like a pure resistance of high value. Thus, no leakage flows. However, the presence of an offset voltage during the ON time will result in an offset error in e_o. This can be removed by introducing a complementary offset later in the system, or by one of the two techniques shown in Fig. 7-9.

In Fig. 7-9(a), diode D_2 is ON during the OFF time of the transistor. The potentiometer can be used to introduce a small voltage equal to the offset voltage during the transistor ON time. The average d-c voltage at the emitter (exclusive of signal voltage) is, therefore, constant.

In Fig. 7-9(b), diode D_2 conducts at the same time as the transistor. In this case, the diode feeds a small reverse current through the transistor. The voltage drop caused by this current opposes the offset voltage. By properly adjusting the potentiometer, enough current can be introduced to cancel the offset.

7-2.2. Drift in Chopper Circuits

There is no guarantee that adequately-compensated chopper circuits will remain so as time passes and temperature changes. However, the effects of drift can also be minimized. When matched devices are being used for complementary compensation, it is reasonable to assume that device processing controls are tight enough that device matching with aging will be adequate.

Drift with temperature can be minimized by proper design. The degree of shift will, in general, be a function of temperature and drive. By testing the device over the desired operating temperature range for several levels of drive,

Sec. 7-3 FETS as Choppers and Switches 215

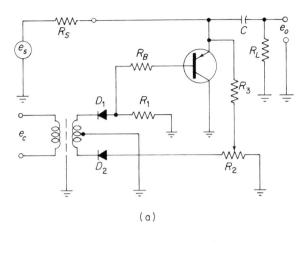

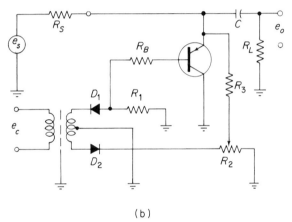

Figure 7-9 Two-junction shunt chopper circuits with offset voltage compensation (Courtesy Motorola)

the drive current for minimum temperature coefficient of offsets at the reference temperature can be determined.

Then, by designing the chopper drive circuitry to match the optimum drive level, minimum temperature drift for the device being used is obtained.

7-3. FETS AS CHOPPERS AND SWITCHES

Field effect transistors make excellent switches and choppers for a multitude of applications, such as modulators, demodulators, sample-and-hold systems, mixing, multiplexing or gating, and many more.

One advantage with the FET is the absence of inherent offset voltage found in two-junction transistors. This is because the conduction path between drain and source is predominantly resistive. In both JFETs and MOSFETs, the conduction channel is either depleted or enhanced by controlling an induced field.

In any type of FET, there is an exceptionally high ratio of OFF resistance to ON resistance for the drain-source channel. The resistance can be as low as several ohms in the ON condition. This results in a very low voltage drop (very low offset voltage). The resistance can be higher than thousands of megohms in the OFF condition, resulting in very little offset current flow.

The extremely high d-c gate input impedance of an FET is also an advantage since little control power is required. In JFETs, the control signal looks into a reverse-biased diode. For MOSFETs, the gate insulation is a high-resistance oxide or nitrite. Thus, the impedance is set by the properties of the insulation layer.

The main limitation of FETs used as switches is the capacitance between the gate and drain, and gate and source. This capacitance feeds through part of the gate control voltage to the signal path. These capacitances are detrimental to high-frequency signal isolation, and also impose a limitation on response times.

7-3.1. FET *Characteristics Applicable to Analog Switches and Choppers*

Figure 7-10 shows the ohmic region of an FET expanded for both positive and negative values of V_{DS}. In the ohmic region, when the FET is fully on, there is a linear relationship between drain current I_D and drain-to-source voltage V_{DS}. The magnitude of this resistance can be changed by varying the gate-to-source voltage V_{GS}. It is in the ohmic region that the FET is useful for both the chopper and analog switch applications.

As shown in Fig. 7-10, to stay in the ohmic region, the drain current must be kept within narrow limits. In other words, a relatively large value of load resistance R_L is required.

Leakage current $I_{D(OFF)}$ versus temperature for a 3N126 (an *N*-channel JFET) is plotted in Fig. 7-11. A very low value of leakage current is important in chopper applications, since this leakage current appears in the output circuit and produces an error voltage.

$I_{D(OFF)}$ versus temperature for a 2N4352 *P*-channel MOSFET is shown in Fig. 7-12. Below 100°C the leakage is so low that accurate readings are more dependent upon available equipment and measurement techniques than on the magnitude of the leakage current. However, the curve of Fig. 7-12 can be projected back to room temperature (shown with dotted lines) and an $I_{D(OFF)}$ of approximately 0.003 pA is read.

Sec. 7-3 FETS as Choppers and Switches

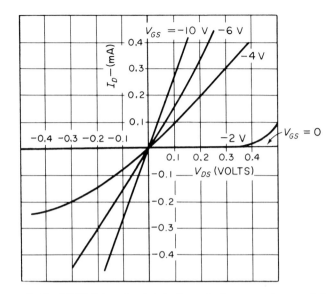

Figure 7-10 2N4352 low-level (ohmic region) output characteristics (Courtesy Motorola)

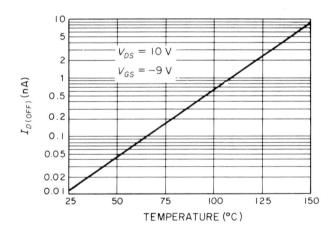

Figure 7-11 3N126 $I_{D(OFF)}$ versus temperature (Courtesy Motorola)

At room temperatures, surface and package leakage account for considerably more than I_{DSS} with a resulting room-temperature leakage of about 0.5 pA. This low leakage current indicates that the OFF voltage error (caused by leakage) will be negligible for most chopper and analog switching circuits. For an enhancement mode MOSFET, $I_{D(OFF)}$ is the same as I_{DSS}.

Drain-to-source resistance R_{DS}, when the FET is ON, is a very important

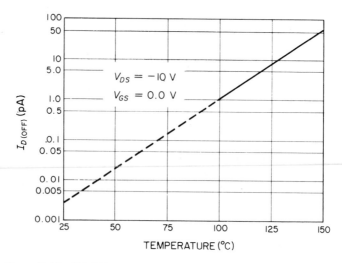

Figure 7-12 2N4352 $I_{D(OFF)}$ versus temperature (Courtesy Motorola)

characteristic in both choppers and analog switches. Figure 7-13 shows R_{DS} versus V_{GS} for three values of temperature for a 2N4352.

On a static basis, there is interest in only two states of the FET; fully ON or fully OFF. The 2N4352 (P-channel MOSFET) needs a negative potential of 10 to 20 V to achieve an R_{DS} minimum. From Fig. 7-13, at $V_{GS} = -10$ V and temperature of 25°C, R_{DS} is 300 Ω.

Compare this with Fig. 7-14 which shows the same characteristics for the MM2102 (an N-channel MOSFET). Note that for $V_{GS} = 10$ V and temperature 25°C, R_{DS} is 100 Ω.

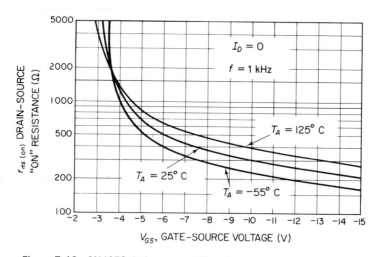

Figure 7-13 2N4352 drain-source ON resistance (Courtesy Motorola)

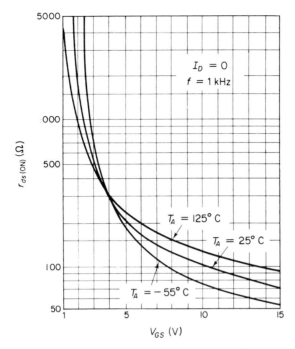

Figure 7-14 MM2102 drain-source ON resistance (Courtesy Motorola)

One reason for this 3:1 improvement in R_{DS} is that in the P-channel device, the carriers are holes, while in the N-channel FET, the carriers are electrons. The mobility of electrons is greater than that of holes, and thus is responsible for part of the improvement in R_{DS}. Since a low R_{DS} is needed in the ON condition, the N-channel MOSFET is preferable for chopper and analog switching applications.

Temperature variations. The variation of FET parameters with temperature can affect operation of a chopper circuit unless allowance is made for such variation in the circuit design. It is important to determine the approximate degree to which each parameter can be expected to change with temperature. Figure 7-15 shows curves of R_{DS}, I_{gss}, and $I_{D(OFF)}$ as a function of temperature for a typical MOSFET.

Capacitance effects. It is very important to know how much capacitance must be charged and discharged during transition times. Figure 7-16 shows a plot of the small-signal, common-source, short-circuit input capacitance C_{iss}, and reverse transfer capacitance C_{rss} versus voltage of the N-channel and P-channel MOSFETs. There is no appreciable change in either capacitance with voltage.

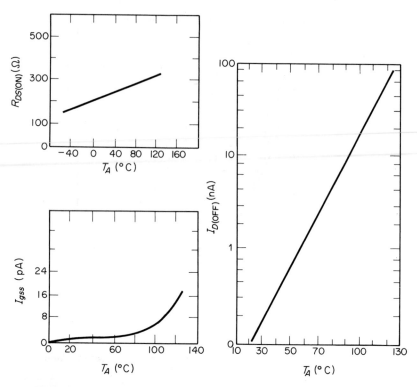

Figure 7-15 R_{DS}, I_{gss}, and $I_{D(OFF)}$ as a function of temperature for a typical MOSFET (Courtesy Motorola)

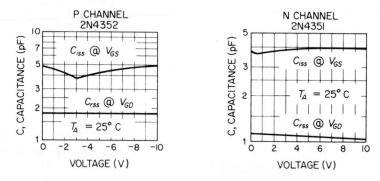

Figure 7-16 C_{iss} and C_{rss} for MOSFETs (Courtesy Motorola)

JFETs are a different case. Figure 7-17 shows that the capacitance of the JFET does vary with voltage (approximately the square root of voltage). Figure 7-17 is a valuable design aid in determining the capacitance at a particular operating point.

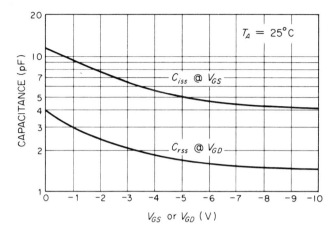

Figure 7-17 3N126 capacitance versus voltage (Courtesy Motorola)

C_{rss} is the capacitance from gate-to-drain, and is the capacitance that causes the feedthrough of the control signal to the load. C_{iss} is the parallel combination of gate-to-drain and gate-to-source capacitance C_{gd} and C_{gs}.

C_{gd} and C_{gs} together form a series capacitance in parallel with R_{DS}. When the FET is used as an analog switch, this capacitance will bypass R_{DS} at high-input frequencies. This bypass will therefore limit the frequency at which the FET can be used as an analog switch.

For example, note that the C_{iss} capacitance is greater than 1 pF at any V_{GS}. At a frequency of 100 MHz, the reactance of a 1-pF capacitor is about 1 kΩ. If the R_{DS} is greater than 1 kΩ, the 100-MHz signals will be bypassed around R_{DS}. If the capacitance is increased to 10 pF, the reactance drops to about 100 Ω. This will bypass most FETs in the OFF condition.

7-3.2. Basic FET Chopper Circuits

Both JFETs and MOSFETs can be used in the three classic chopper configurations: series, shunt, and series-shunt.

The series chopper is the most commonly used. The basic circuit, equivalent circuit, and equations for the series chopper are shown in Fig. 7-18. In order to operate in the ohmic region when the FET is ON, the drain current must be limited to a low value. Generally, a large value of load resistance R_L is used to limit the current. In the case of a series chopper, a high value of load resistance also minimizes the ON voltage error due to R_{DS}.

The leakage of an FET (and in particular the MOSFET) is low, and the resultant OFF voltage error is small. Typically, for a MOSFET chopper with R_L at 100 kΩ, the OFF error is less than a microvolt at room temperature.

The shunt chopper, shown in Fig. 7-19, performs the chopping function by

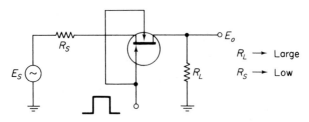

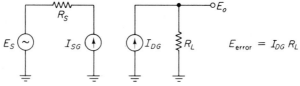

Figure 7-18 FET series chopper equivalent circuits (Courtesy Motorola)

periodically shorting the input to ground. From the ON equivalent circuit, Fig. 7-19(b), note that the shunt circuit is advantageous where a large source resistance R_S is present.

The series-shunt chopper, shown in Fig. 7-20, operates on the following principle: When Q_1 is ON, Q_2 is OFF; conversely, when Q_1 is OFF, Q_2 is ON. The equivalent ON-OFF circuits are also shown in Fig. 7-20. When Q_1 is ON and Q_2 is OFF, the output is similar to that of the series chopper, except for the small error introduced by the leakage of Q_2.

When Q_1 is OFF and Q_2 is ON, the OFF voltage error due to the leakage of Q_1 is reduced since R_{DS2} appears in parallel with R_L. This can be seen from the OFF voltage error equation.

However, the leakage current of a MOSFET is quite small, so the series-shunt circuit can not be justified for the sake of minimizing the error due to

FET SHUNT CHOPPER

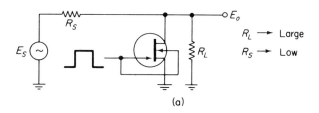

(a)

EQUIVALENT ON CIRCUIT

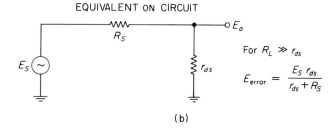

For $R_L \gg r_{ds}$

$$E_{error} = \frac{E_S \, r_{ds}}{r_{ds} + R_S}$$

(b)

EQUIVALENT OFF CIRCUIT

$$E_{error} = (I_{DGO} R_S + E_S) \frac{R_L}{R_L + R_S}$$

(c)

Figure 7-19 FET shunt chopper equivalent circuits (Courtesy Motorola)

leakage. The series-shunt circuit does have a definite advantage in the area of high-frequency chopping.

In the simple series chopper, when the FET is OFF, C_{rss} must be discharged through the load resistor R_L. The relatively long-time constant ($C_{rss}R_L$) will limit the chopping frequency. In the series-shunt chopper, however, every time the series device turns OFF, the shunting device is turned ON, and the low resistance of Q_2 will parallel R_L. The RC time constant will, therefore, be greatly reduced, and the chopping frequency can be increased significantly.

7-3.3. Effect of Loads on MOSFET Choppers

Operation of all MOSFET chopper circuits is greatly affected by the magnitude of the source and load resistances, R_S and R_L. Figure 7-21 lists the output voltages of the three basic chopper circuits for various com-

FET SERIES–SHUNT CHOPPER

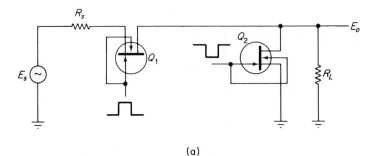

(a)

EQUIVALENT ON CIRCUIT

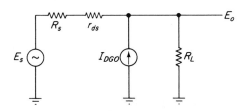

$$E_{error} = \frac{R_L [E_S + I_{DGO}(R_S + r_{ds})]}{R_L + R_S + r_{ds}}$$

(b)

EQUIVALENT OFF CIRCUIT

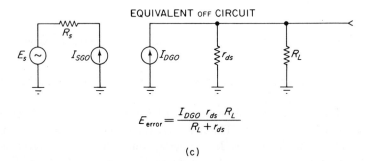

$$E_{error} = \frac{I_{DGO}\, r_{ds}\, R_L}{R_L + r_{ds}}$$

(c)

Figure 7-20 FET series-shunt chopper equivalent circuits (Courtesy Motorola)

binations of source and load resistances. It is assumed that the input voltage E_S is 1 mV, and the drain-to-source resistance R_{DS} is 100 Ω in the ON condition, and 1000 MΩ in the OFF condition. The gate leakage resistance (typically 10^{12} Ω, or more) is neglected. The following conclusions can be drawn from the data shown in Fig. 7-21.

Sec. 7-3 FETS as Choppers and Switches 225

Approximate Output Voltage e_o (μV)
(Max output 1 mV)

Source Resistance R_S (Ω)	Load Resistance R_L (Ω)	Shunt Chopper ON	Shunt Chopper OFF	Series Chopper ON	Series Chopper OFF	Series-Shunt Chopper ON	Series-Shunt Chopper OFF
1 M	1 M	0.1	500	500	1	500	0.0001
100 k	1 M	1	900	900	1	900	0.0001
100	1 M	500	1000	1000	1	1000	0.0001
0	1 M	1000	1000	1000	1	1000	0.0001
1 M	100 k	0.1	90	90	0.1	90	0.0001
1 M	100	0.05	0.1	0.1	0.0001	0.1	0.00005
100 k	100 k	1	500	500	0.1	500	0.0001
100	100	333	500	333	0.0001	333	0.00005

Figure 7-21 Steady-state chopper output voltage for various source and load resistances (Courtesy RCA)

1. Only the series or series-shunt circuit should be used when R_S is less than $R_{DS(ON)}$.
2. In general, R_L should be high. In any event, R_L should be much greater than $R_{DS(ON)}$.
3. R_L should always be greater than R_S.
4. Performance of the series-shunt circuit is equal to or better than that of either the series or shunt chopper alone, for any combination of R_S and R_L.

7-3.4. Effect of Interelectrode Capacitance on MOSFET Choppers

The interelectrode capacitances of MOSFET choppers have their greatest effect as frequency increases. The capacitances are of little concern at low frequencies. The high-frequency effect of the capacitances is shown in Fig. 7-22, which is the a-c equivalent of a MOSFET shunt chopper.

The input capacitance C_{gs} increases the rise time of the gate driving signal, and thus increases the switching time of the chopper. However, this effect is not usually a serious limitation because the switching time of the MOSFET depends primarily on the input and output time constants. Switching times as short as 10 nS can be achieved when a MOSFET is driven from a low-impedance source and the load resistance is less than about 2 kΩ.

The output capacitance C_{ds} also tends to limit the maximum frequency that can be chopped. When the reactance of this capacitance becomes much lower than the load resistance R_L, the chopper becomes ineffective because XC_{ds} is essentially in parallel with R_L and $R_{DS(OFF)}$.

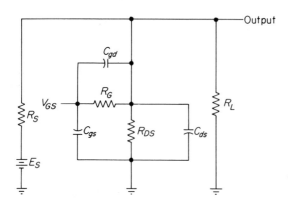

Figure 7-22 a-c equivalent circuit of MOSFET shunt chopper (Courtesy RCA)

The feedthrough capacitance C_{gd} is the most important of the three interelectrode capacitances because it couples a portion of the gate drive signal into the load circuit, and causes a *voltage spike* to appear across R_L each time the gate drive signal changes state. C_{gd} and R_L form a differentiating network which allows the leading edge of the gate drive signal to pass through. The output capacitance C_{ds} is beneficial to the extent that it helps reduce the amplitude of the feedthrough spike.

The effect of the feedthrough spikes can be reduced by several methods. Typical approaches include the following

1. Use of a clipping network on the output when the input signal to be chopped is fixed in amplitude.
2. Use of a low chopping frequency.
3. Use of a MOSFET that has a low feedthrough capacitance (typically a fraction of a picofarad).
4. Use of a gate drive signal that has poor rise and fall times (sloppy square wave).
5. Use a source and load resistance as low as feasible. Of course, low values of R_S and R_L produce greater error voltages, as discussed in Sec. 7-3.3.
6. Use of shield between the gate and drain leads.
7. Use of a series-shunt chopper circuit.

7-3.5. Practical JFET Choppers

Figure 7-23 shows a simple, but complete JFET series chopping circuit. The maximum chopping frequency is about 200 kHz. The limitation is primarily due to the long RC time constant for discharging $C_{rss}C_{gs}$ through the 10 kΩ resistors. The time constant can be shortened, and the frequency increased, if the 10 kΩ resistor is reduced. However, this results in a greater voltage drop across the divider consisting of the drain-source resistance and the load resistor.

Sec. 7-3 FETS as Choppers and Switches 227

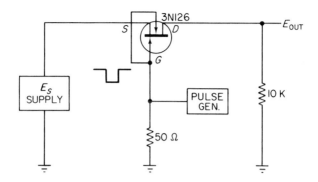

MAXIMUM CHOPPING FREQUENCY $f_{(max)} \simeq 200$ kHz
MAXIMUM INPUT VOLTAGE $E_{S(max)} \simeq +2$ V, -0.4 V

Figure 7-23 Practical series chopper using N-channel JFET (Courtesy Motorola)

The maximum allowable input voltages are $+2$ V and -0.4 V. The reason for these voltage limitations is the ON-OFF requirement of the JFET. For the particular N-channel JFET shown (3N126), the ON condition is 0 V, and the OFF conditions is a V_{GS} greater than 4 V. The ON condition is obtained by grounding the gate. When the source (E_S supply) starts to go positive, there is a negative potential from gate to source. The negative gate-source voltage causes R_{DS} to increase, and the FET starts to turn OFF.

When the input (source) goes more negative than -0.4 V, the P-N diode from gate to source becomes forward-biased, and the FET starts to turn ON. Hence, the negative limitation of -0.4 V.

In addition to these positive and negative input voltage limitations, there is a minimum input voltage limitation due to feedthrough spikes in the output channel. This feedthrough is caused by C_{rss} and C_{iss}, as previously discussed. For inputs less than about 10 mV, the feedthrough spikes become an appreciable part of the output waveform (particularly at high frequencies).

There are several circuit techniques useful for minimizing these spikes. First, the control signal (pulse generator) at the gate can be a sloppy square wave. That is, the dv/dt of the input pulse should be kept as low a value as possible. Sharp corners on the waveform should be avoided. (A sine wave could be used.) Next, a capacitor can be connected across the output to filter the spikes. Finally, if a fixed amplitude output is acceptable, a clipper circuit can be connected across the output.

Modified series chopper for large input voltages. When an FET is OFF, large negative input voltages may tend to turn the FET ON again. The circuit of Fig. 7-23 can be modified to get around this input voltage limitation. For

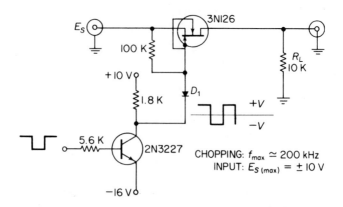

Figure 7-24 JFET circuit for large input voltages (Courtesy Motorola)

example, in order to overcome the limitation of maximum input voltage when the FET is turned ON, a circuit similar to that of Fig. 7-24 can be used.

With the Fig. 7-24 circuit, the FET is ON, the driver transistor is OFF, and a potential of $+10$ V appears at diode D_1. This reverse-biases D_1. When the input (source) goes positive, the 100-kΩ resistor from source to gate makes the gate follow in potential as long as the diode does not become forward-biased. If the diode D_1 starts to conduct, a negative potential appears from gate to source, and the FET starts to turn OFF. For negative input voltages, there is no problem as before, since the gate will follow the source until D_1 avalanches.

The circuit of Fig. 7-24 also improves the performance when the FET is OFF. Under these conditions, the driver transistor is turned ON, and when the gate of the FET is at -15 V, the FET will remain OFF for inputs up to -10 V.

The maximum input voltage of ± 10 V is a function of the bias voltage. The real limitation for the input voltage to this circuit is the source-to-gate breakdown voltage. Typical breakdown voltage is 50 V for a JFET (and about 30 V for a MOSFET). With different values of bias, the input could be increased to about $+22$ V (when both positive and negative inputs are applied), or to about $+44$ V (when only positive inputs are used. Approximately -6 V is required to keep the 3N126 OFF).

7-3.6. Practical MOSFET Choppers

Figure 7-25 shows a practical MOSFET chopper. Note that this circuit is similar to that of the JFET circuit in Fig. 7-23. Operation of the two circuits is similar. However, since a MOSFET does not have a P-N junction at the input, the characteristics of the two circuits are somewhat different.

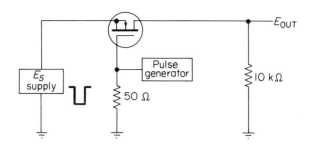

Maximum chopping frequency $F_{max} \approx 300$ kHz
Maximum input voltage $E_{S(max)} \approx +3$ V -0.4 V

Figure 7-25 Practical series chopper using N-channel MOSFET

The maximum chopping frequency of the MOSFET circuit (Fig. 7-25) is about 300 kHz, depending on the capacitances of the MOSFET. This limitation is primarily due to the long RC time constant for discharging C_{rss} through the 10 kΩ resistor. The time constant can be shortened and the frequency increased if the 10 kΩ resistor is reduced. However, this results in a greater error voltage.

The maximum allowable input voltages are set by the V_{GS} limits of the MOSFET. In addition, there is a minimum input voltage limitation, due to feedthrough spikes in the output channel. This feedthrough is caused by C_{rss} and C_{iss}, as previously discussed. For inputs less than about 10 mV, the feedthrough spikes become an appreciable part of the output waveform (particularly at high frequencies).

As discussed, there are several circuit techniques useful for minimizing these spikes. The techniques include sloppy input or control signals, capacitors across the output, and (if a fixed amplitude output is acceptable) a clipper at the output.

Practical MOSFET series-shunt chopper for high-frequency use. Figure 7-26 is a series-shunt high-frequency chopper using *complementary* enchancement-mode MOSFETs. Using the components shown, the circuit will operate satisfactorily at frequencies up to about 5 MHz. An N-channel and a P-channel MOSFET are used as the series and shunting devices, respectively. This allows one drive circuit for both devices.

When the series MOSFET is OFF, the R_{DS} of the shunting MOSFET is about 200 Ω. This value parallels R_L to ground, and reduces the net output load resistance to about 200 Ω (the parallel combination of 200 Ω and 10 kΩ). Thus, the RC time constant is reduced to 2 per cent of its original value.

The circuit of Fig. 7-26 can also be modified to accept large values of input voltage, as described in Sec. 7-3.5 (for the circuit of Fig. 7-24).

230   Transistor Switches   Chap. 7

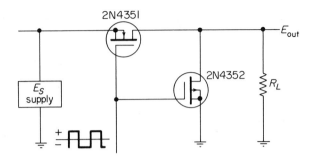

Maximum chopping frequency $F_{max} \approx 5$ MHz
Maximum input voltage $E_{S(max)} \approx +0.5$ V, -4 V

Figure 7-26 Series-shunt chopper for high-frequency applications using complementary enhancement mode MOSFETs (Courtesy Motorola)

Practical MOSFET series-shunt chopper for low-input voltages. A series-shunt chopper capable of low-level chopping is shown in Fig. 7-27. Two N-channel MOSFETs *with matched* C_{rss} are used in the circuit. The gate drives for the pair are produced by a high-current type (or current-mode type) astable multivibrator. Refer to Sec. 3-2 of Chapter 3.

The high-current mode multivibrator drive results in good frequency stability. In addition, the complementary outputs are not delayed in time, with respect to each other. That is, when one output is turning OFF, the other output must be concurrently turning ON.

By matching C_{rss}, the feedthrough spikes in the output of the chopper can almost be eliminated. Complete elimination of feedthrough spikes is difficult to obtain since the turn-on and turn-off characteristics of MOSFETs are not symmetrical.

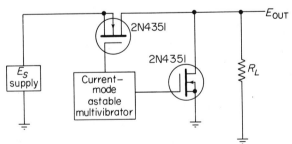

Maximum chopping frequency $F_{max} \approx 5$ MHz
Maximum input voltage $E_{S(min)} \approx \pm 10$ μV

Figure 7-27 Series-shunt chopper for low-input voltages (Courtesy Motorola)

Sec. 7-3   FETS as Choppers and Switches   231

Dual-gate MOSFET chopper circuits. The circuits shown in Figs. 7-28 and 7-29 use dual-gate MOSFETs in chopper or gating circuits.

In the shunt choppers of Fig. 7-28, the MOSFET is normally conductive so e_o is low. A negative gating pulse turns off the MOSFET so that approximately 50 per cent of e_g appears at the output terminals. The circuit of Fig. 7-28(a) features an addition control potential V_{G2}. A d-c potential may be applied to the second gate, thereby establishing the value of desire channel ON resistance, or $R_{DS(ON)}$. As an alternative, the second gate can function as a "coincidence-gate" to reduce e_o to a low value. This requires a positive-going pulse applied to gate 2 simultaneously with a positive-pulse to gate 1.

The circuits shown in Fig. 7-29 function in a manner opposite to those of

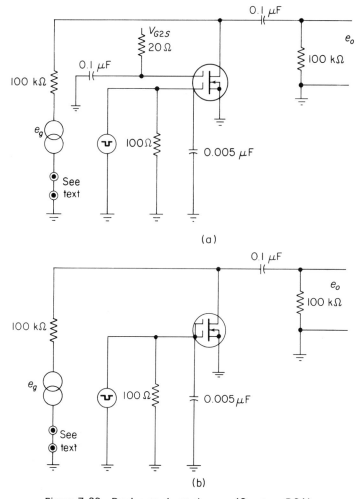

Figure 7-28 Dual-gate shunt chopper (Courtesy RCA)

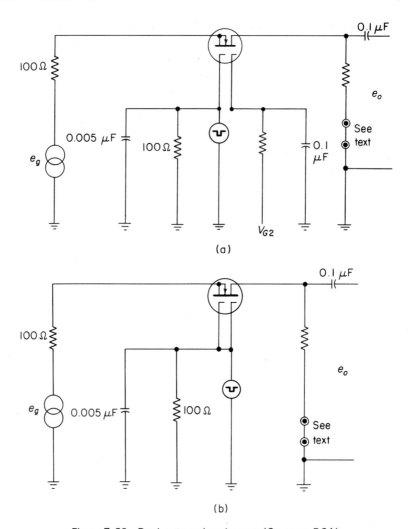

Figure 7-29 Dual-gate series chopper (Courtesy RCA)

Fig. 7-28. That is, output voltage appears at e_o in the *absence of a gating signal*. Consequently, a negative gating signal reduces the level of e_o. The dual-gate configuration can be made into an *OR*-circuit. That is, a negative signal applied to gate 2 of sufficient magnitude to override V_{G2} will also reduce the level at e_o.

The circuits of Figs. 7-28 and 7-29 show a jumper connected between two terminals in the drain-to-ground-return circuits. The circuits assume a peak generator level e_g of less than 0.2 V. Should the signal exceed this value, it is possible that the parasitic "diode" between the drain and semiconductor substrate will be driven into conduction and load the signal. This can be overcome by connecting a suitable d-c potential in lieu of the jumper, so that a positive potential is applied to the drain. The magnitude of this voltage should equal or exceed the peak value of the signal from e_g.

7-3.7. FET *Analog Switching Circuits*

Most of the information described thus far for FET choppers can be applied to FET analog switches. By definition, the analog switch is a device that either transmits an analog signal without distortion, or completely blocks it off.

Typically, an N-channel JFET is capable of passing input frequencies up to 20 MHz, without appreciable distortion or attenuation.

The frequency response curves for a 3N126 JFET used as an analog switching device are shown in Fig. 7-30. In running these curves, the JFET was operated as a switch between a sine-wave input voltage of 0.2 V, and an *RF* voltmeter monitoring the output. Readings of the output voltage were taken for two values of load resistance R_L. The output was measured both with FET ON and OFF.

For a load resistance $R_L = 10$ kΩ and the FET ON, the curve shows the output to be 2.5 dB down at 20 MHz. The $R_L = 10$-kΩ curve for the OFF condition shows the output to be 22 dB at 20 MHz, or approximately 20 dB of separation between ON and OFF.

With a 50-Ω load, considerable loss is experienced due to the R_{DS} (approximately 500 Ω) of the FET. The output is 14 dB down at 20 MHz for the

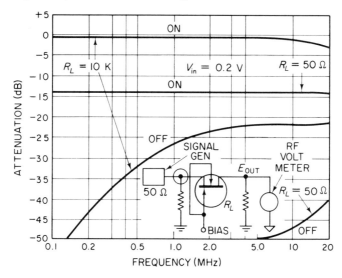

Figure 7-30 Frequency response of 3N126 (Courtesy Motorola)

FET turned ON, and 40 dB down for the FET turned OFF, resulting in a separation of 26 dB.

The circuit of Fig. 7-24 can be used as an analog switch as well as a chopper. Such a switch is able to pass an input signal of ± 10 V, with a frequency up to 20 MHz, without appreciable distortion or attenuation.

Another analog switching circuit using a MOSFET is shown in Fig. 7-31. (Of course, the circuit can also be used as a chopper.) The problem of handling large positive or negative values of input voltage is solved here in a somewhat different fashion than for the JFET circuit of Fig. 7-24.

With the MOSFET circuit of Fig. 7-31, there is no P-N junction to worry about since the gate is insulated from the rest of the FET. This eliminates the need for a diode at the gate. There are, however, P-N junctions from substrate-to-source, and substrate-to-drain. These junctions must not be allowed to become forward-biased. One way to accomplish this is to cut off the substrate lead and leave it floating. However, since the substrate is connected to the package (metal can), this result in quite a bit of pick-up.

A more practical solution is shown in Fig. 7-31. Here, the substrate-to-source junction is coupled with a diode. In turn, the substrate is coupled back to ground through a 100-kΩ resistor.

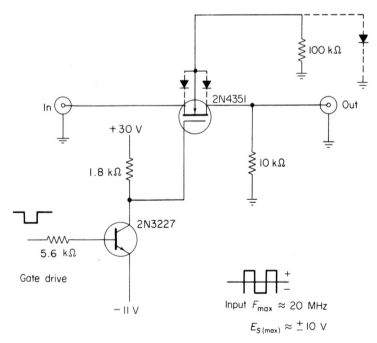

Figure 7-31 MOSFET analog switch for large input voltages (Courtesy Motorola)

Sec. 7-4   Transistor Inverters and Converters   235

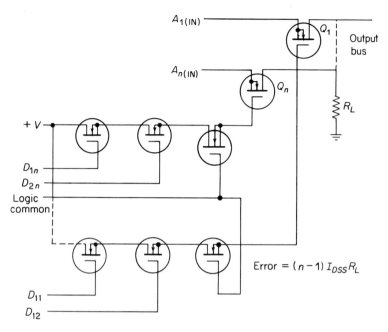

Figure 7-32 Commutator network using N-channel MOSFETs (Courtesy Motorola)

Commutator using FET analog switches. The analog switch can be used in a commutator circuit, as shown in Fig. 7-32. Each switch has a three-input AND gate in series with the gate drive. In order to turn a switch ON, a positive potential is required at the gate. To accomplish this, the three inputs of the AND gate must be "true" (at a positive potential).

Assume that A_n is to be sampled. Suppose the logic common is a clock signal, that D_{1n} is an order from the control system to sample A_n, and D_{2n} is a ready signal from the device to be sampled. When all of these conditions are true at the same time, switch Q_n is turned ON and A_n is sampled. The circuit of Fig. 7-32 can be modified to accept large values of input voltage.

In this type of commutating circuitry, where only one channel is turned ON at a time, an error is introduced due to the leakage of the other FET switches. Assuming that the leakage is the same for all switches, an approximation of the error signal is given by $(n - 1)I_{DSS}R_L$.

7-4. TRANSISTOR INVERTERS AND CONVERTERS

Transistor inverters and converters are often used in solid-state power supplies. A transistor inverter is a solid-state circuit used to change a d-c input (generally from a battery) into an a-c output. If the a-c output is again converted back to d-c (generally at a higher voltage) the circuit is called a converter.

There are many inverter/converter circuits in common use. The most common are circuits using two transistors, and one or two transformers. In addition, there are bridge-type inverters that use several transistors (typically four) and several transformers (or one transformer with multiple windings). Further, there are special-purpose inverter/converter circuits with such features as resistive-coupled transistors, saturable inductors, multivibrator drive, and single transistors with single transformers.

All of these circuits are discussed in this section. We shall concentrate on design fundamentals for the most popular circuits. Practical information as to the selection or design of transformers is given at the end of the section.

7-4.1. Basic Transistor Inverter Theory

The operational theory of a typical transistor inverter can best be understood by reference to Fig. 7-33. The circuit shown is that of a two-transistor single-transformer inverter using common-emitter connections. The transformer B-H curve illustrates the relationship between magnetizing force (H) and flux density (B). The horizontal or H line shows magnetizing force (or current flow through the winding). The vertical or B line shows flux density. Note that as force (current) increases, flux density also increases, up to the saturation points (L-K and M-J). Further increases in current produces no further increase in flux density. Typical waveforms for the circuit are also included in Fig. 7-33. Note that transistor saturation voltage $V_{CE(\text{sat})}$ is not to be confused with transformer saturation current.

Assume that transistor Q_1 is nonconducting, Q_2 is conducting, and the transformer is saturated at point J on the B-H curve.

When Q_1 starts to conduct, the voltage developed across the primary windings N_1 induces voltage in the feedback windings N_3 such as to rapidly drive Q_1 into saturation, and turn Q_2 off. When this transition is completed, a constant voltage V_P is applied to N_1. Voltage V_P is the value of V_{CC}, less the value of $V_{CE(\text{sat})}$, or $V_P = V_{CC} - V_{CE(\text{sat})}$. With a constant V_P applied, the flux increases at an approximately constant rate from point J to point K on the B-H curve.

As long as the transformer core remains nonsaturated, magnetization current is small. As saturation (point K) is approached, high magnetization current is required to keep the flux constant. When the reflected load current (from N_2), plus this sharply increasing magnetization current exceeds the collector current which Q_1 can supply, Q_1 begins to come out of saturation. This causes V_P to decrease, and Q_1 to be cutoff, ending the half-cycle.

As flux in the transformer core collapses from point K to point B_R, voltage is induced in the winding which biases transistor Q_2 into conduction

Sec. 7-4   Transistor Inverters and Converters

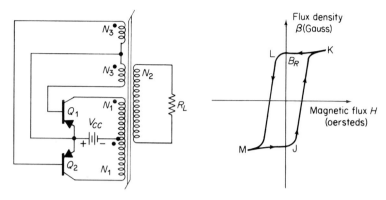

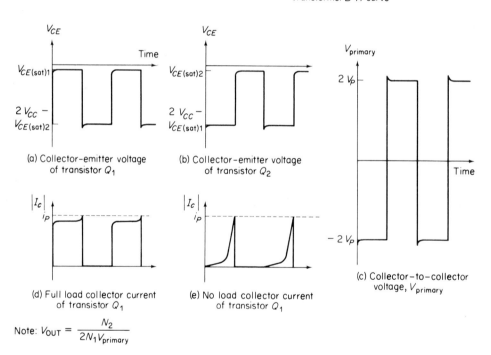

Note: $V_{OUT} = \dfrac{N_2}{2N_1 V_{primary}}$

Figure 7-33   Basic one-transformer inverter circuit and waveforms (Courtesy Motorola)

and initiates the next half-cycle. The operation is similar to the first half-cycle except that V_P is applied to the other half of the primary, causing a reversal of polarity in the induced output-voltage. Transistor Q_2 conducts until the core is driven into negative saturation at point M on the *B-H* curve. As flux collapses from M to J, the full-cycle is completed.

It can be seen from the collector-to-emitter voltage waveforms that each transistor is subjected (in the off condition) to a voltage *approximately twice the supply voltage*, plus any induced voltage that may occur in the circuit due to leakage inductance, etc. Also, the *same maximum collector current* i_p is required for switching action whether this current is primarily reflected load current (Fig. 7-33(d)), or totally magnetized current (Fig. 7-33(e)). This condition will obviously limit efficiency at low-output loads.

Operation frequency of the inverter is determined by the voltage V_P, and by the saturation characteristics of the transformer core, according to the relationship

$$\text{Frequency (in hertz)} \approx \frac{V_p 10^8}{4B_s A N_1}$$

where:
B_s is saturated flux density in gauss
A is cross-sectional area of the core in square centimeters (cm²)
N_1 is the number of turns on one-half of the primary

7-4.2. Basic Common-Collector and Common-Base Inverters

Figure 7-34 shows the basic common-base transistor inverter, using two transistors and one transformer. The characteristics are essentially the same as for the common-emitter circuit of Fig. 7-33. However, the common-base circuit of Fig. 7-34 is used where the supply voltage is very low. A major drawback of the Fig. 7-34 circuit is that the feedback windings N_3 must carry the high current of the emitters.

Figure 7-35 shows two versions of the common-collector transistor inverter. These circuits are used for high-power PNP transistors where the collector is connected to the case, and the case is mounted directly on a heat sink. In the circuit of Fig. 7-35(a), the feedback N_3 windings must be much larger than the emitter windings N_1. The circuit of Fig. 7-35(b) uses an autotrans-

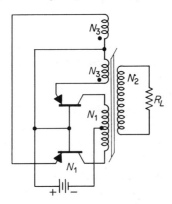

Figure 7-34 Basic common-base inverter circuit (Courtesy Motorola)

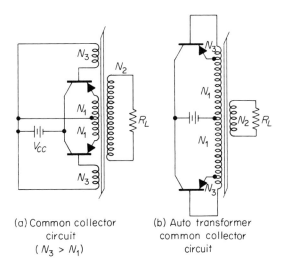

(a) Common collector circuit ($N_3 > N_1$)

(b) Auto transformer common collector circuit

Figure 7-35 Basic common-collector inverter circuit (Courtesy Motorola)

former connection. That is, the transformer has a single primary winding, but the winding is provided with feedback taps.

It should be noted that the common-collector and common-base inverters are not in general use.

7-4.3. Basic Two-Transformer Inverters

Figure 7-36 shows two versions of typical two-transformer inverter designs. Such designs are used for applications where high-output power and higher operating frequencies are involved. Two-transformer designs also permit frequency control, and more efficient transformation of the output voltage.

Operation of the two-transformer inverters is similar to that of the one-transformer design, except that in each two-transformer circuit only a small feedback transformer (T_2 of Fig. 7-36) need be saturated. Output current is handled by T_1. Since magnetization current of the smaller T_2 is low, high-current levels due to transformer saturation currents are reduced significantly, when compared to one-transformer design. Likewise, the transistors need not carry the high transformer saturation currents.

Compare the collector current characteristics of one-transformer (Fig. 7-33(d) and 7-33(e)) design with those of two-transformer (Fig. 7-36(c) and 7-36(d)) circuits. With the circuits of Fig. 7-33, the collector current remains high with both no-load and full-load, indicating that the currents in one-transformer designs are largely transformer saturation currents. In two-

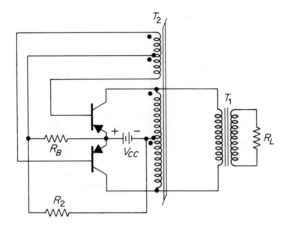

(a) Two-transformer inverter with simple output transformer T_1

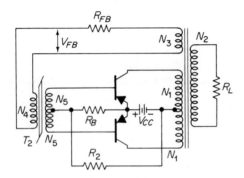

(b) Two-transformer inverter having easily regulated V_{FB}

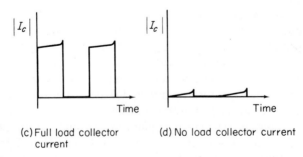

(c) Full load collector current

(d) No load collector current

Figure 7-36 Typical two-transformer inverter designs and waveforms (Courtesy Motorola)

transistor designs, the required current drops to near zero when the load is removed.

In the circuit in Fig. 7-36(a), the large output transformer T_1 uses a conventional core material, and the small control or feedback transformer T_2 uses a special core. Use of a conventional output transformer T_1 with normal core material permits lower transformer costs, as well as higher efficiency.

In the circuit of Fig. 7-36(b), the frequency is determined (primarily) by the feedback voltage V_{FB}. This feedback voltage can be regulated to provide a constant frequency, or can be changed (by means of a variable R_{FB}) to provide variable frequency.

7-4.4. Basic Single-Transistor Inverter

Figure 7-37 shows a basic one-transistor inverter circuit. Such circuits are useful in low-power inverter applications where lower efficiency is of secondary importance when compared to lower initial cost. In the circuit of Fig. 7-37, positive feedback, transformer saturation, and switching mechanism are similar to the two-transistor inverter, except that resetting action is caused by capacitor C, rather than by a second transistor in push-pull configuration.

In effect, the circuit of Fig. 7-37 is a single-transistor power oscillator. For proper operation, R_L and C must not overload the oscillator during the saturating half-cycle, so that transistor conduction will be maintained until the transformer core saturates. Likewise, care must be taken to protect the transistor against excessive voltage "backswing" when it is turned off.

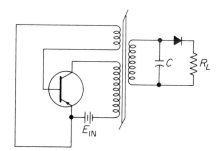

Figure 7-37 Basic one-transistor inverter circuit

7-4.5. Basic Driven Inverters

Figure 7-38 shows the diagram of a basic driven inverter. In such circuits, the output power transistor switching is accomplished by a multivibrator drive circuit, rather than by feedback from the output transformer.

Multivibrator drive transistor inverters are used mostly in precision systems requiring carefully controlled frequency, waveform, etc., and for load

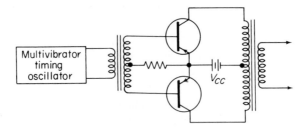

Figure 7-38 Typical driven inverter

independent systems. Load independent systems are especially attractive in the case of reactive loads, and when transient or starting conditions impose loads which would cause self-oscillating inverters to shut down or operate abnormally.

The power requirements of the extra multivibrator circuits are largely offset by nonsaturation of the output transformer and elimination of the feedback drive. The multivibrator driven inverter is not inherently less efficient than a self-oscillating inverter. However, use of the driven power stage as a linear amplifier, rather than a saturated switch, will result in high dissipation in the transistors, and in low system efficiency.

For these reasons, the driven inverter is generally used only where precision frequency control is required. About the only other application is for highly reactive loads.

7-4.6. Basic Resistive-Coupled Inverter

Figure 7-39 shows the basic resistive-coupled transistor inverter. The circuit obtains the required squarewave drive from the transistor output by cross-coupled resistor feedback. There are two major drawbacks to

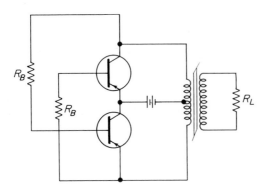

Figure 7-39 Resistive-coupled transistor inverter

the Fig. 7-39 circuit. First, the circuit is not efficient due to losses in the feedback resistors R_B. Second, the frequency of the inverter is difficult to set and maintain.

7-4.7. Basic Saturable Base Inductor Inverter

Figure 7-40 shows the basic saturable base inductor inverter. This circuit is essentially a simplification of the two-transformer inverter, except that the saturating transformer has been replaced by a saturable inductor connected between the bases of the two transistors. This permits the output transformer to have a normal core and normal saturation characteristics.

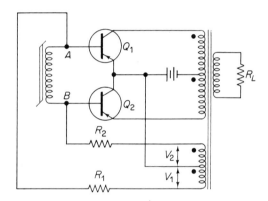

Figure 7-40 Saturable base inductor transistor inverter

Operation of the Fig. 7-40 circuit is as follows. With transistor Q_1 conducting, Q_1 is driven on by base current I_B. This base current is approximately equal to $(V_1 - V_{BE})/R_1$. Transistor Q_2 is biased off by V_2 through R_2.

The voltage across the inductor is initially V_{AB}, which is approximately equal to $V_2 + V_{BE}$. As the inductor saturates, V_{AB} drops, essentially shorting out the feedback circuit. When this occurs, Q_1 loses its drive and turns off.

The magnetization current of the transformer then reverses the voltage, drives the system over in the opposite sense and brings Q_2 toward conduction. As soon as the inductor comes out of saturation, the positive feedback is effective and the second half-cycle is begun.

Saturable base inductor inverter design is relatively simple, being largely determined by the following relationships: $V_1 = V_2$, $R_1 = R_2$, $I_{B1} = I_{B2} = (V_1 - V_{BE})/R_1$, and operating frequency (in hertz) is determined by

$$\frac{(V_1 + V_{BE})10^8}{4NAB_s}$$

where:

> N is the number of turns in the saturable inductor
> A is the cross-sectional area of the saturable inductor core in square centimeters (cm²)
> B_s is the saturated flux density (in gauss) of the saturable inductor

7-4.8. Basic Bridge-Type Inverters

Figure 7-41 shows three versions of the basic bridge-type inverter. Bridge-type inverters are generally used when high-input supply voltage exceeds transistor voltage capabilities. Bridge-type inverters were popular some years ago when transistors generally could not be operated at high voltages. In the circuits of Fig. 7-41, the transistors are never subjected to voltages greater than the supply voltage. This is unlike the simple push-pull circuits where transistors could be subjected to an instantaneous voltage of almost twice the supply. The reduction of voltage applies to all circuits in Fig. 7-41. However, in the half-bridge circuit in Fig. 7-41(c), transistor current must double to maintain the original output-power.

Bridge-type inverters have one serious drawback which must be considered. This has to do with the current-voltage excursions of the transistors as the circuit switches. If the previously non-conducting side of the circuit turns on before the other side is essentially off, high voltage and high current may be imposed on the transistors, and the maximum transistor voltage-current limits may be exceeded. Also, high transient voltages may be generated.

These problems may be somewhat alleviated by reducing transistor "on" drive, by transistor protection against transients, or by compensating base-drive networks which retard turn-on of the non-conducting device. One possibility is to use a driven bridge having the input waveforms of Fig. 7-41(d). However, because of these problems, and because present-day transistors can withstand relatively high voltages, bridge-type inverters have become less popular.

7-4.9. Basic Series Inverters

Figure 7-42 shows a basic series inverter. The circuit is essentially a number of two-transistor inverter circuits connected in series to one common transformer. The purpose of this circuit is the same as for bridge-type inverters (to reduce maximum voltage drop across transistors). This is done by dividing the supply voltage equally among several inverters. In the circuit in Fig. 7-42, the voltage is divided between two basic inverters. In theory, any number of inverters can be used. Under such conditions, each device is required to withstand a voltage approximately equal to: $2V_{cc}$/number of inverters.

Sec. 7-4 Transistor Inverters and Converters 245

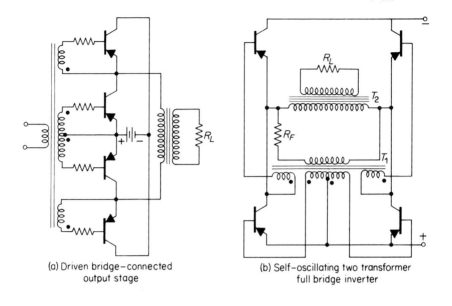

(a) Driven bridge-connected output stage

(b) Self-oscillating two transformer full bridge inverter

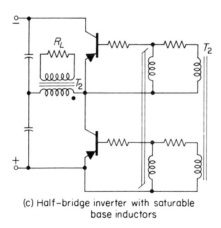

(c) Half-bridge inverter with saturable base inductors

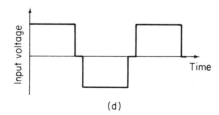

(d)

Figure 7-41 Typical bridge inverter circuits (Courtesy Motorola)

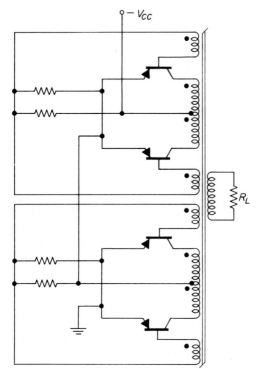

Figure 7-42 Basic series inverter (Courtesy Motorola)

The circuit of Fig. 7-42 has some obvious limitations. A special transformer must be designed. Likewise, the circuits and transformer must be such that the voltage will divide equally among the series stages. If not, the output waveform will be abnormal and efficiency will drop.

7-4.10. Basic Inverter Specifications

The following is a list of specifications commonly applied to inverters. As is the case with any specification, the more characteristics simultaneously required and the tighter the tolerances, the more difficult and costly the design will be.

In the absence of specifications, inverters are generally designed for a given input and output, *with maximum efficiency*. That is, the output should be maximum for a given input.

Input voltage: range and nominal
Output power
Output voltage
Output frequency accuracy
Regulation of output voltage and frequency vs. load and input voltage
Load power factor

Sec. 7-4 Transistor Inverters and Converters 247

Output waveform

Harmonic distortion of output, if sinusoidal, vs. load, power factor, input voltage

Overall efficiency vs. loading

Operating environments (temperature, etc.)

Size and weight

Protection required (as against shorted output, reversed polarity input, etc.)

7-4.11. Inverter Starting Circuits

In general, the basic circuits discussed thus far will not oscillate readily unless some means is provided to start oscillation. This is especially true at full load and low temperature, the most severe starting requirements for resistive loads. The discussion of basic inverter operation assumed that one of the transistors was conducting. The function of the starting circuit is to assure this condition.

A simple, commonly used starting circuit is shown in Fig. 7-43(a). In this circuit, R_1 and R_2 form a simple voltage divider to bias the transistors to conduction before oscillation starts.

A good guideline for the base starting bias developed by the circuit of Fig. 7-43(a) is to use 0.3 V for germanium transistors, and 0.5 V for silicon. The base bias voltage V_B can be found by

$$V_B = \frac{R_1 V_{CC}}{R_1 + R_2}$$

Since R_1 occurs in the feedback circuit in series with the base of each circuit half, R_1 must not exceed

$$R_B = \frac{V_{FB} - V_{EB}}{I_B}$$

If R_1 is set equal to R_B, then R_1 and R_2 are uniquely determined for any given starting bias. The value of R_2 may be adjusted if starting characteristics are not satisfactory. This straight forward starting technique in Fig. 7-43(a) is advantageous in that only resistor components need be added to the circuit, but has the disadvantage of additional power dissipation (which may become excessive in high power circuits).

An improved but somewhat more costly starting circuit using a diode is shown in Fig. 7-43(b). This diode circuit requires less dissipation than its resistance counterpart and is less temperature dependent. Operation is similar to the circuit in Fig. 7-43(a), except that when power is applied to the bases in the Fig. 7-43(b) circuit, the transistors are driven negative by full supply voltage and oscillation starts rapidly.

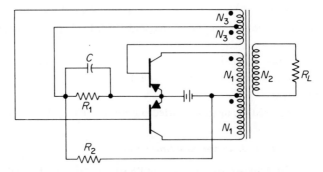

(a) Simple resistive self-starting circuit with speed-up capacitor across R_1

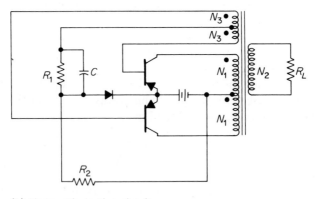

(b) Diode self-starting circuit

Figure 7-43 Starting circuits for transistor inverters (Courtesy Motorola)

Inverter loads, such as capacitive filters, starting motors, or incandescent lamps may temporarily present extremely high loads during the starting period. Starting requirements of such loads are often somewhat simplified by using a driven inverter. The driven inverter may be preferable to the circuit complications needed to assure self-oscillation.

7-4.12. Inverter Speed-Up Circuits

Inverter speed-up circuits improve transistor switching and inverter efficiency. Improved switching is especially important at higher frequencies, as is discussed in later paragraphs. A common speed-up method is to add a capacitor to the circuit. The circuits of Fig. 7-44 usually produce improved switching waveforms.

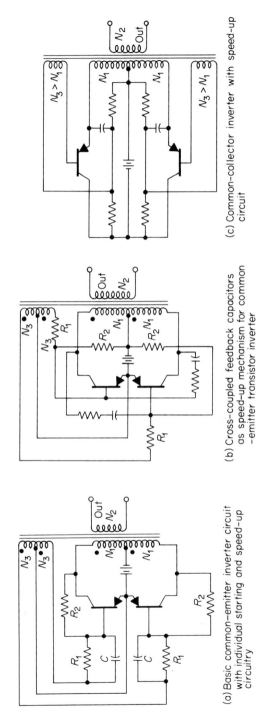

Figure 7-44 Speed-up circuits for transistor inverters (Courtesy Motorola)

7-4.13. Special Inverter Design Considerations

In many cases, it must be determined whether circuit modifications are desirable and necessary to protect against inverter damage due to output overload or short circuit, input transients, input polarity reversal, etc.

Shorted output causes cessation of inverter oscillation, but intermediate overload may cause transistor failure.

Undesirable collector-emitter voltage spikes caused by input voltage transients or high transformer leakage inductance may be prevented by Zener diodes connected collector-to-emitter. Despiking may also be accomplished by a series resistor and capacitor across the full primary winding or between collector and base of each transistor, but these arrangements slow transistor switching.

If the transistors will not sustain reverse voltages resulting from input polarity reversal, protection may be provided by a diode in series with the input terminals. Such an arrangement is shown in Fig. 7-45.

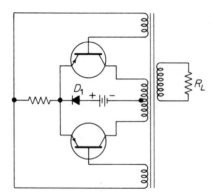

Figure 7-45 Reverse polarity protection by means of diode D_1 in series with input circuit of inverter

7-4.14. Selecting Inverter Components

The critical components in any inverter or converter design are the transistors and transformers. The transformers may be off-the-shelf or special design. The following notes summarize important features to be considered in selecting transistors and inverters.

Transistor selection. The transistor type selected must be capable of maximum collector current i_p where

$$i_p = \frac{P_{\text{IN}}}{V_{CC}} = \frac{P_o}{(V_{CC}) \times \text{overall inverter efficiency}}$$

The transistor must have collector-emitter breakdown in excess of the "off" voltage, which is approximately twice the supply voltage V_{CC}. Also, a margin of safety should be applied to allow for voltage transients from leakage inductance, transients of input voltage, etc. Transistor surge rating *three times* supply voltage is a reasonable guideline. The transistor must have a sufficiently safe operating area such that operating load lines are well within transistor capabilities.

Inverter transistor power losses limit overall inverter efficiency. Most transistor power losses occur during switching. This is shown in Fig. 7-46 which illustrates typical inverter transistor waveforms for I_c, V_{CE}, and $P_C(I_c V_{CE})$. Switching load lines for inverter transistors and switching power losses are summarized in Fig. 7-47.

Inverter transistor efficiency is maximumized by high gain, low $V_{CE(sat)}$, fast switching response, and low leakage in the off condition. Likewise, inverter transistors must be capable of the required power dissipation, and must have a breakdown voltage BV_{EBO} high enough to sustain feedback reverse bias (or a clamped emitter-base reverse voltage). Unfortunately, each of these factors is achieved at some expense to the others, so a tradeoff must be considered.

Transformer selection or design. Inverter transformers can be off-the-shelf components or special designs. In later paragraphs, we describe (in practical terms) some commercial inverter transformers with regards to winding data, etc. Here, we shall discuss theoretical approaches to inverter transformer design.

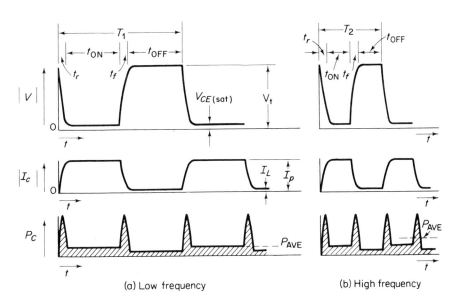

Figure 7-46 Typical inverter waveforms for I_C, V_{CE}, $P_C(= I_C \times V_{CE})$ for transistors of d-c to a-c inverters (Courtesy Motorola)

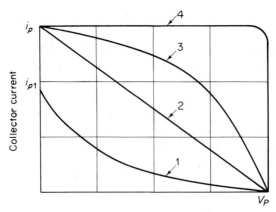

Collector current / Collector emitter voltage

Typical Switching Load Lines for Transistors

Switching losses for the above load lines are given by the following relationships.
  t_s = switching time (rise time or fall)
  f = frequency
1. $P_1 = 2/9\, V_1\, i_{p1}\, t_s\, f$ where peak power occurs at $t_s/3$
2. $P_2 = 1/6\, V_p\, i_p\, t_s\, f$ where peak power occurs at $t_s/2$
3. $P_3 = 2/9\, V_p\, i_p\, t_s\, f$ where peak power occurs at $t_s/3$
4. $P_4 = V_p\, i_p\, t_s\, f$

Figure 7-47 Summary of transistor switching power loss (Courtesy Motorola)

In a one-transformer inverter, the transformer determines frequency as well as output and feedback voltages. To accomplish this with acceptable transformer efficiency is the essence of the inverter transformer design problem.

The basic inverter frequency equation of Sec. 7-4.1 can be rearranged to find the number of turns for N_1 as follows

$$N_1 = \left(\frac{V_P \times 10^8}{4F}\right)\left(\frac{1}{B_s A}\right)$$

In this equation, the cross-sectional area A of the core includes allowance for the stacking factor of laminated or tape-wound cores. Core selection is somewhat arbitrary, but must allow enough turns for good coupling and enough window area for the required windings. Also, core selection should provide for minimum loss. As a guideline, core losses should be comparable in magnitude to resistive losses of the windings. For example, if winding resistance losses are 2 per cent, core losses should not exceed 2 per cent. From a practical standpoint, core choice is influenced by cost, size, and weight of the overall transformer.

Tape wound toroids of 50-50 nickel-iron are generally considered as best choices for inverter transformer cores. This material has high B_s, a square hysteresis loop, low core loss, and is little affected by temperature over the useful temperature range of transistors. (The term "square hysteresis loop" can best be understood by reference to Fig. 7-33. If the flux density versus magnetizing force diagram were perfectly square, it would indicate that flux density increased in direct proportion to force. In practical core materials, a perfectly square loop is not possible since density is never quite directly proportional to force.) The use of toroid windings is recommended since toroids (when properly wound) give very close coupling. This is very desirable in inverter transformers. C-cores and E-, U-, and I-cores of good square loop material are also popular.

Once a tentative core selection has been made, the number of turns for N_1, N_2, and N_3 can be found as follows

$$N_1 = \frac{V_P \times 10^8}{4B_s AF} \qquad N_2 = \frac{k_1 V_o N_1}{V_P} \qquad N_3 = \frac{k_2 V_{FB} N_1}{V_P}$$

where k_1 and k_2 are multipliers to compensate for transformer voltage drops and losses. As a first trial, let $k_1 = k_2 = 1.05$ to 1.1.

In determining V_o for required output power, it should be noted that for a square wave $V_o = V_{peak} = V_{rms}$.

V_{FB} must be greater than transistor V_{EB} in the "on" condition, but must be less than BV_{EBO}, unless transistor reverse V_{BE} is clamped. With V_{FB} approximately equal to V_{EB}, the circuit performance is highly dependent upon transistor and temperature.

High V_{FB} and a series base resistor R_B reduces sensitivity to V_{EB} and frequently improves transistor turn-off, especially if a speed-up capacitor is used across R_B (See. 7-4.12). However, losses in R_B are directly proportional to V_{FB}. The first trial value for R_B can be found by

$$R_B = \frac{V_{FB} - V_{EB}}{I_B}$$

As a guideline for power inverters, a V_{FB} on the order of 3 V will give adequate base drive stability and control, without exceeding reasonable dissipation losses.

The winding factor is another consideration in transformer core design. The ratio of wire area to available window area is the winding factor. A winding factor of about 0.4 is considered typical for a toroidal core. The factor is increased to 0.7 or 0.8 for cores with rectangular windows.

Wire size is usually calculated on the basis of 700 to 1000 circular mils per ampere. Keep in mind that duty cycle should be applied when calculating winding currents. For example, using the circuit of Fig. 7-33, the N_1 and N_3 windings have 50 per cent duty cycles, whereas the N_2 winding has a 100 per cent duty cycle. Also, in calculating total winding areas, it must be remembered that there are two each N_1 and N_3 windings.

Transformer efficiency is found by

$$\text{Efficiency} = \frac{(\text{watts out}) \times 100}{(\text{watts out}) + (\text{core losses}) + (\text{copper losses})}$$

Transformer voltage regulation is found by

$$\frac{I_2\left[R_S + \left(\frac{N_2}{N_1}\right)^2 R_p\right]}{E_2} = \frac{\left[R_S + \left(\frac{N_2}{N_1}\right)^2 R_p\right]}{R_L}$$

where R_p and R_S are resistances of primary and secondary windings.

Several design attempts may be needed to achieve a core and winding combination that will provide the required frequency and voltage relationships with acceptable efficiency. Commercial transformers for both one-transformer and two-transformer inverters are available for a variety of standard supply voltage, output voltage, and output power combinations. These are discussed in Sec. 7-4.15.

7-4.15. Practical Transistor Inverter Design

Figures 7-48, 7-49 and 7-50 are working schematics for three converters. These converters operate with d-c inputs of 12 V and 28 V, and produce d-c output voltages from 100 V to 500 V. Output current ratings are from 1 A to 25 A. Output voltages can be increased or decreased by changing the number of secondary turns. If desired, the circuits of Figs. 7-48 through 7-50 can be used as inverters by omitting the rectifier diodes and filter capacitors at the output windings.

Note that components are assigned values, type numbers, or winding data. The values for these components depend upon the desired output. Proper selection of these values, based on the design considerations and examples of this section, will provide the desired output power. Other power outputs can be produced by means of intermediate values. However, the tabulated outputs should provide the designer with sufficient choice for most applications. The values tabulated are the nearest standard parts available that will meet or exceed the minimum specifications required for the particular converter.

Sec. 7-4 Transistor Inverters and Converters 255

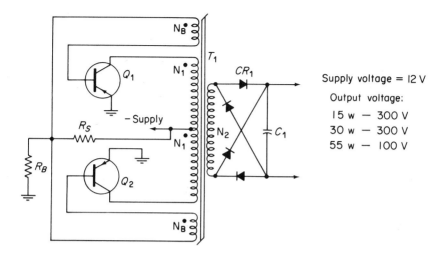

Power rating (watts)	Q_1 and Q_2	R_B (Ω)	R_S (Ω)	CR_1	C_1 (μF)	N_1 turns AWG	N_2 turns AWG	N_B turns AWG	T_1 type
15	2N1038	15	1200	1N2071	2	70 #18	1800 #30	20 #30	Arnold #5772D2
30	2N1042	15	1500	1N2071	4	78 #16	2000 #29	30 #29	Magnetics #500172A
55	2N456	5	180	1N2071	6	29 #17	275 #24	6 #24	Magnetics #500352A

Figure 7-48 DC-DC converter (15 to 55 W) (Courtesy Texas Instruments)

Following are brief descriptions of the various converters to show the relationship of circuit function to design problem.

One transformer converter. When the required power outputs are between 15 and 55 W, the circuit of Fig. 7-48 is used. Note that only one transformer is required. Any voltage imbalance in the circuit causes one of the transistors, for example Q_1, to conduct a small amount of current. Regeneration turns Q_2 off, and Q_1 is driven into saturation. The collector current increases. As the core becomes saturated, the collector current increases rapidly and is limited only by the resistance in the collector circuit and the transistor characteristics.

When the core is saturated, the induced voltage in the winding is zero. The resulting lack of base drive causes Q_1 to turn off, and the collector current drops to zero. The drop in collector current causes the polarity in all wind-

256 Transistor Switches Chap. 7

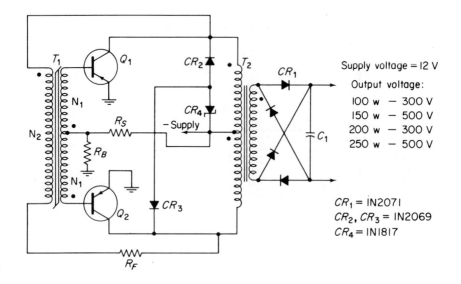

Supply voltage = 12 V
Output voltage:
100 w — 300 V
150 w — 500 V
200 w — 300 V
250 w — 500 V

CR_1 = IN2071
CR_2, CR_3 = IN2069
CR_4 = IN1817

Power rating (watts)	Q_1 and Q_2	R_B (Ω)	R_S (Ω)	R_F (Ω)	C_1 μF	N_1 turns AWG	N_2 turns AWG	T_1 type	T_2 part no. (see text)
100	2N511	2	100	5	10	48 #24	185 #28	Magnetics #500942A	440402-1
150	2N512	2	75	10	20	48 #22	185 #26	Magnetics #501812A	440404-1
200	2N513	1	75	5	20	35 #20	140 #26	Magnetics #500262A	440406-1
250	2N514	1	75	5	30	35 #20	140 #24	Magnetics #500262A	440408-1

Figure 7-49 DC-DC converter (100 to 250 W) (Courtesy Texas Instruments)

ings to reverse, therby biasing Q_1 off and Q_2 on. When the core approaches negative saturation, Q_2 is turned off, and the current drops to zero, turning Q_1 on.

Resistors R_B and R_S are added to place a bias on the bases of both Q_1 and Q_2. This bias provides a starting current and reduces the effect of base-emitter voltage variations. The frequency of oscillation is determined by the design of transformer T_1.

The secondary output voltage is rectified by the bridge-connected diodes CR_1, and filtered by capacitor C_1.

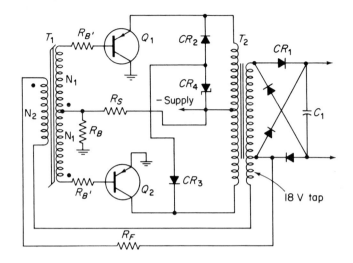

Supply voltage = 28 V   Output voltage = 300 V
                         Output power = 500 w

CR_1 = IN1126 Q_1, Q_2 = 2N514A R_S = 75 Ω
CR_2, CR_3 = IN2069 R_B = 0.5 Ω R_F = 10 Ω
CR_4 = IN1825 $R_{B'}$ = 0.5 Ω C_1 = 40 μF

T_1 = Arnold #523302
N_1 = 35 turns #20 AWG
N_2 = 140 turns #24 AWG
T_2 = 440413 −1 (see text)

Figure 7-50 DC-DC converter (500 W) (Courtesy Texas Instruments)

Two transformer converters. When the required power output is in excess of about 55 W, two transformers are used, as shown in Figs. 7-49 and 7-50. However, only transformer T_1 has a saturable core. Thus, the extra current necessary at saturation is small compared to the load current. This allows use of a small drive transformer T_1 to drive a much larger, relatively inexpensive, power transformer T_2 which steps up the output voltage to the required value.

When one of the transistors (for example Q_1) conducts, the collector swings from the supply voltage to near zero (saturation). The voltage builds up across the primary of T_2 and is applied to the primary of T_1 through feedback resistor R_F. Transistor Q_2 is biased off, whereas Q_1 is biased into conduction. As soon as the core of T_1 reaches saturation, the increasing current causes additional voltage drop across R_F, thus aiding the regenerative action (Q_2 off; Q_1 on). Transistor Q_2 continues in this state until reverse saturation of the transformer is reached. The circuit then switches back to the initial state, and the cycle is completed.

The collector current of the conducting transistor rises to the value of the load current, plus the magnetizing current of T_2, and the feedback current needed to produce the drive. The magnetizing current of T_2 is never more than a fraction of the rated load current, since T_2 is not allowed to saturate.

Resistors R_B and R_S are added to place a bias on the bases of Q_1 and Q_2. This bias provides a starting current and reduces the effect of base-emitter voltage variations. The frequency of oscillation is determined by the design of transformer T_1 and the value of feedback resistor R_F.

7-4.16. Design Considerations

The prime consideration in any converter is that the circuit will produce the desired output voltage, at a given current, or into a given load resistance. The next consideration is the efficiency of the converter (output power versus input power). Frequency of oscillation may also be a consideration in some applications.

Figure 7-51 through 7-58 show graphs of output voltage, output power, frequency, and efficiency, versus load current. As shown, the output voltage regulation is less than 7 per cent from half to full-rated current. The output power is almost a straight line function of load current. The frequency of the circuit in Fig. 7-48 decreases slightly under load. The frequency of the circuits in Figs. 7-49 and 7-50 remain almost constant under load. The efficiency of all converters is greater than 80 per cent at rated load.

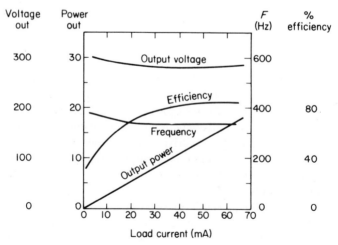

Voltage in = 12 V

Figure 7-51 15-W converter characteristics (Courtesy Texas Instruments)

Sec. 7-4   Transistor Inverters and Converters   259

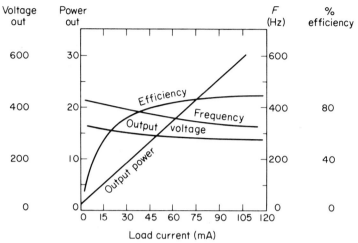

Figure 7-52 30-W converter characteristics (Courtesy Texas Instruments)

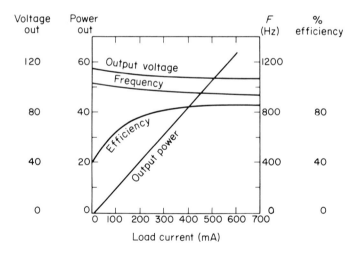

Figure 7-53 55-W converter characteristics (Courtesy Texas Instruments)

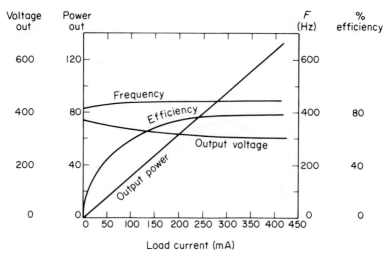

Figure 7-54 100-W converter characteristics (Courtesy Texas Instruments)

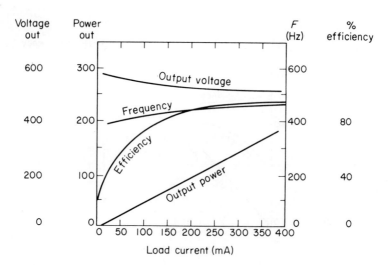

Figure 7-55 150-W converter characteristics (Courtesy Texas Instruments)

Sec. 7-4   Transistor Inverters and Converters

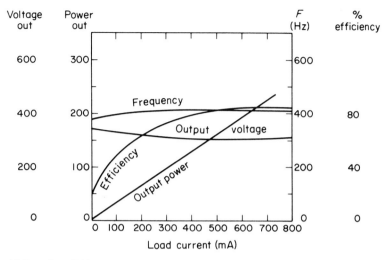

Figure 7-56   200-W converter characteristics (Courtesy Texas Instruments)

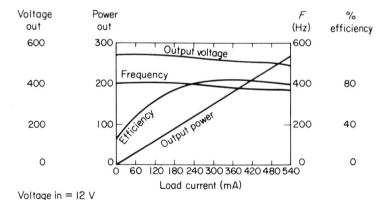

Figure 7-57   250-W converter characteristics (Courtesy Texas Instruments)

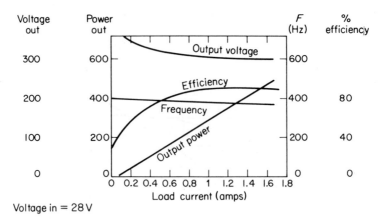

Figure 7-58 500-W converter characteristics (Courtesy Texas Instruments)

Use of the these graphs, in conjunction with the schematic and tabulated values, is best understood by reference to the following design example.

7-4.17. Design Example

Assume that the circuit in Fig. 7-49 is to provide 200 W of output power. The output voltage must be 300 V, with 12-V input. The frequency is 400 Hz.

Converter efficiency. A converter efficiency of 80 per cent is assumed (from the graph in Fig. 7-56). The necessary input power is

$$\text{Power input} = \frac{\text{power output}}{\text{efficiency}} = \frac{200}{0.8} = 250 \text{ W}.$$

Collector current. The collector current of each transistor is given by

$$\text{Collector current} = \frac{\text{power input}}{\text{supply voltage}} = \frac{250}{12} = 20.8 \text{ A}$$

Transistor characteristics. The transistors must be selected on the basis of maximum collector breakdown voltage and maximum collector current (20.8 A, or greater, for safety). Each transistor, during its off time, is subjected to approximately twice the supply voltage or (2 × 12) 24 V. The recommended transistor type is 2N513, as shown in Fig. 7-49.

Base current. The base current of each transistor is

$$\text{Base current} = \frac{\text{collector current}}{\text{minimum } \beta} = \frac{20.8}{20} = 1.04 \text{ A}$$

Note that minimum β (at the rated collector current) is obtained from the 2N513 datasheet.

Base drive voltage. The base-emitter drive voltage is made equal to twice the maximum base-emitter voltage shown on the transistor datasheet. This reduces the effect of variation in base-emitter voltage among transistors. The maximum base-emitter voltage for a 2N513 is 2 V. Thus, the base-emitter drive voltage is 4 V.

Base drive power. The power required for base drive is given by

$$\text{Base drive power} = 4 \times 1.04 = 4.16 \text{ W}$$

Drive transformer T_1 efficiency and power input. A 90 per cent efficiency is assumed for the drive transformer T_1. Thus, the power supplied to the primary of T_1 is

$$\text{Primary drive power} = \frac{\text{base drive power}}{\text{efficiency}} = \frac{4.16}{0.9} = 4.6 \text{ W}$$

Turns ratio. The turns ratio of the drive transformer is chosen as 4:1. Since the required base drive voltage is 4 V (at the secondary of T_1) the primary voltage of T_1 is 16 V.

Drive transformer T_1 primary current. With a primary power input (primary drive power) of 4.6 W, and a primary voltage of 16 V, the primary current of T_1 is 4.6/16 or 287 mA.

The design considerations discussed thus far establish the characteristics necessary to order (or construct) drive transformer T_1. These characteristics are available with the transformer tabulated in Fig. 7-49.

Power transformer T_2 characteristics. The characteristics for ordering (or constructing) transformer T_2 are given in Fig. 7-59. The following notes should also be considered. The input impedance of T_2 (at the operating frequency) should be approximately 40 times the load resistance. The core loss at the rated load is approximately 5 per cent of the total power output. The primary and secondary winding losses are approximately 1 per cent of the total power output.

Base resistors. The value of R_B is chosen to drop approximately half of

264   Transistor Switches   Chap. 7

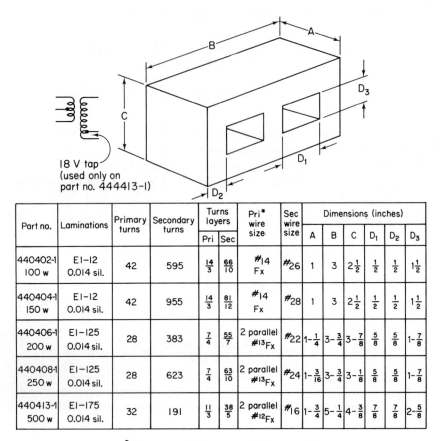

Part no.	Laminations	Primary turns	Secondary turns	Turns layers Pri	Turns layers Sec	Pri* wire size	Sec wire size	Dimensions (inches) A	B	C	D_1	D_2	D_3
440402-1 100 w	EI-12 0.014 sil.	42	595	$\frac{14}{3}$	$\frac{66}{10}$	#14 Fx	#26	1	3	$2\frac{1}{2}$	$\frac{1}{2}$	$\frac{1}{2}$	$1\frac{1}{2}$
440404-1 150 w	EI-12 0.014 sil.	42	955	$\frac{14}{3}$	$\frac{81}{12}$	#14 Fx	#28	1	3	$2\frac{1}{2}$	$\frac{1}{2}$	$\frac{1}{2}$	$1\frac{1}{2}$
440406-1 200 w	EI-125 0.014 sil.	28	383	$\frac{7}{4}$	$\frac{55}{7}$	2 parallel #13 Fx	#22	$1-\frac{1}{4}$	$3-\frac{3}{4}$	$3-\frac{7}{8}$	$\frac{5}{8}$	$\frac{5}{8}$	$1-\frac{7}{8}$
440408-1 250 w	EI-125 0.014 sil.	28	623	$\frac{7}{4}$	$\frac{63}{10}$	2 parallel #13 Fx	#24	$1-\frac{3}{16}$	$3-\frac{3}{4}$	$3-\frac{1}{8}$	$\frac{5}{8}$	$\frac{5}{8}$	$1-\frac{7}{8}$
440413-1 500 w	EI-175 0.014 sil.	32	191	$\frac{11}{3}$	$\frac{38}{5}$	2 parallel #12 Fx	#16	$1-\frac{3}{4}$	$5-\frac{1}{4}$	$4-\frac{3}{8}$	$\frac{7}{8}$	$\frac{7}{8}$	$2-\frac{5}{8}$

*Primary parallel wires are bifilar wound

Figure 7-59   Schematic and transformer core for T_2

the 4-V drive voltage. The value of R_S is made as large as possible to provide some starting current, yet keep losses at a minimum.

Feedback resistor. The value of R_F is chosen to drop the necessary voltage to give 16 V across the primary of T_1, at the rated load.

8 OSCILLATOR CIRCUIT DESIGN

Virtually all of the classic vacuum tube oscillator circuit designs can be duplicated with transistors. In addition, there are a number of circuits where transistors are superior. Likewise, certain types of transistors are particularly well suited to specific oscillator circuits. In this chapter, we shall discuss the design of those transistor oscillator circuits which have proved their value over the years.

The main concern in any oscillator design is that the transistor will oscillate at the desired frequency and will produce the desired voltage or power without damage. Most transistor oscillator circuits operate with power outputs of less than 1 W. Many transistors will handle this power dissipation without heat sinks.

8-1. *LC* AND CRYSTAL-CONTROLLED OSCILLATORS

LC oscillators are those that use inductances (coils) and capacitors as the frequency-determining components. Typically, the coils and capacitors are connected in series- or parallel-resonant circuits, and adjusted to the desired operating frequency. Either the coil or capacitor can be variable. *LC* oscillators are used at higher frequencies (RF). Both the classic Hartley and Colpitts oscillators can be implemented with transistors. However, the Colpitts is generally the most popular.

Transistor *LC* oscillators can be crystal-controlled. That is, a quartz crystal can be used to set the frequency of operation, with an adjustable *LC* circuit used to "trim" the oscillator output to an exact frequency. In addition

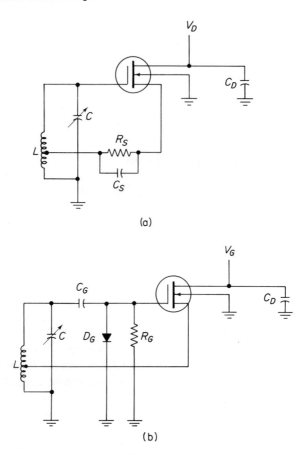

Figure 8-1   Basic MOSFET Hartley oscillator circuits

to the Hartley and Colpitts, there are a number of other crystal oscillator circuits suitable for transistors. These include the Pierce oscillator, harmonic or overtone oscillators, and oscillators that use two transistors to provide the necessary feedback required for oscillation.

8-1.1. Basic Transistor LC Oscillators

Figure 8-1 shows two arrangements of the Hartley oscillator circuit. A MOSFET is shown. However, the circuit can also be implemented with a JFET or a two-junction transistor. UJTs are not generally suitable for LC oscillators.

The circuit of Fig. 8-1(a) uses a bypassed source resistor to provide proper operating conditions. The circuit of Fig. 8-1(b) uses a gate-leak resistor and biasing diode. The amount of feedback in either circuit is dependent on the position of the tap on the coil. Too little feedback results in a feedback signal voltage at the gate insufficient to sustain oscillation. Too much feedback causes the impedance between source and drain to become so low that the circuit is unstable. Output from the Hartley circuits can be obtained through inductive coupling to the coil, or through capacitive coupling to the gate.

One problem common to all oscillators is the *class of operation*. If an oscillator is biased class A (with some current flowing at all times), the output waveform will be free from distortion, but the circuit will not be efficient. That is, power output will be low in relation to power input.

For the purposes of calculation, input power for a transistor oscillator can be considered as the product of collector (or drain) current and voltage. Class A oscillators are usually not good for RF and are generally limited to those applications where a good waveform is the prime consideration.

A class C oscillator (where current is cutoff by feedback) is far more efficient. This cuts both power and heat requirements. At radio frequencies, the waveform is usually not critical, so class C is in common use for RF circuits.

One drawback to class C is that the oscillator may not start in the reverse-bias condition. This can be overcome by forward biasing the transistor to start the oscillator (start collector or drain current flow). The arrangement can be aided by an unbypassed-emitter (or unbypassed-source) resistor. Collector or drain current flow will build up a reverse bias across the resistor.

A particular problem with this bias scheme is that too much reverse bias may cause the transistor to cut off during the "on" half-cycle. A "variable bias" can be used to overcome this problem and maintain correct bias relationships.

With two-junction transistors and JFETs, the variable bias charge is obtained by rectifying part of the oscillator signal and filtering the bias, using a large value capacitor. With such transistors, the base-emitter or gate-source "diode" junction serves as the rectifier, with the coupling (feedback) capacitor serving as the bias filter (to retain the correct bias charge during the "on" cycle). With MOSFETs, no "diode" exists, so external components are used, as shown in Fig. 8-1.

Because the bias is variable (changes with the amplitude of the oscillator signal), the capacitor charge must also change. If the capacitor is too small, the oscillator may not start easily, or there will be distortion. If the capacitor is too large, the charge changes slowly, and the oscillator operates intermittently as a blocking oscillator.

To sum up, if the selected transistor is capable of producing the required power at the operating frequency, and the correct component values are selected for the resonant circuits, the only major problem in oscillator design is the correct bias point. Often, this must be found by trial-and-error test of the circuit in breadboard form.

In the design examples described in later paragraphs of this section, the class of operation is set by the amount of feedback, rather than the bias point. That is, the transistor is biased for an optimum operating point, and then feedback is adjusted for the desired class of operation.

Figure 8-2 shows a MOSFET used in two forms of the Colpitts oscillator circuit. Again, either two-junction transistors or JFETs can be used. Colpitts circuits are more commonly used in VHF and UHF equipment than the Hartley circuits because of the mechanical difficulty involved in making the tapped coils required at high frequencies. Feedback is controlled in the Colpitts oscillator by the ratio of capacitance C' to C''.

LC radio frequency oscillators require resonant circuits for their operation. Therefore, all of the design considerations of Chapter 6 apply to the resonant circuits for LC oscillators. Likewise, the design considerations of Chapter 6 should be studied when RF oscillators are to be used with frequency multiplier and/or power amplifiers.

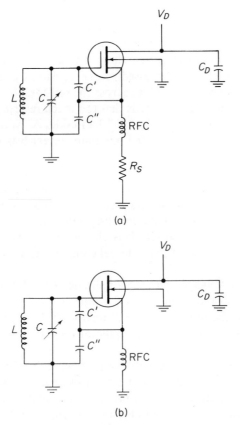

Figure 8-2 Basic MOSFET Colpitts oscillator circuits

Sec. 8-1 *LC* and Crystal-Controlled Oscillators 269

8-1.2. Basic Transistor Crystal-Controlled Oscillator

Transistors operate officiently in *crystal oscillator* circuits such as the Pierce-type oscillator shown in Fig. 8-3. The Pierce-type oscillator is very popular because of its simplicity and minimum number of components. No *LC* circuits are required for frequency control. Instead, the frequency is set by the crystal.

At frequencies below 2 MHz, a capacitive voltage divider may be required across the crystal. The connection between the voltage-divider capacitors must be grounded so that the voltage developed across the capacitors is 180° phase inverted.

It is frequently desirable to operate crystals in communications equipment at their *harmonic or overtone frequencies.* Figure 8-4 shows two circuits designed for overtone operation. Additional feedback is obtained for the

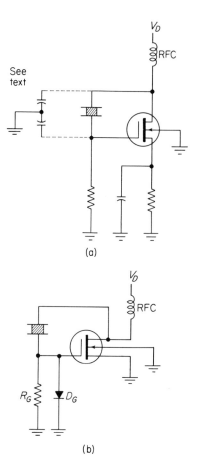

Figure 8-3 Basic MOSFET Pierce-type oscillator circuits

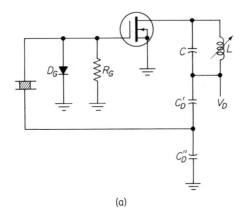

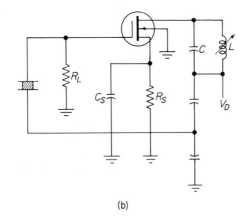

Figure 8-4 MOSFET crystal oscillator circuits permitting operation at overtone or harmonic frequencies

overtone crystal by means of a capacitive divider as the *LC* circuit (or "tank") bypass. Most third-overtone crystals operate satisfactorily without this additional feedback, but for the 5th and 7th harmonics this extra feedback is required. The *LC* tank in Fig. 8-4 is not fully bypassed and thus produces a voltage that aids oscillation. The crystals in both circuits are connected to the junction of the two capacitors C'_D and C''_D. The ratio of these capacitors should be approximately 1:3.

The circuit of Fig. 8-5 operates well at *low frequencies*. The crystal is located in the feedback circuit between the sources of the two transistors and operates in the series mode. Capacitor C_2 is used for precise adjustment of the oscillator frequency. A reduction in C_2 capacitance increases the frequency slightly.

The practical limit of crystal fundamental resonance is about 25 MHz for transistor crystal oscillators. From 20 to 60 MHz, 3rd overtone crystals are used. 5th overtone crystals are used for frequencies between 60 and 120 MHz,

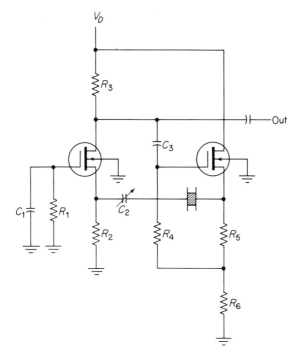

Figure 8-5 Low-frequency MOSFET crystal oscillator

and 7th overtone crystals at frequencies above 120 MHz. Usually, 150 MHz is the top limit for a transistor crystal oscillator. Multipliers are used above this frequency.

In most transistor circuits, the crystal series-resonant frequency is most appropriate. At "overlapping" frequencies, where either a 5th or 7th overtone crystal can be used (such as 115 to 125 MHz), the power output is about the same for either crystal. Of course, the 7th overtone may present other problems, such as more critical adjustments and a tendency to oscillate at the 5th overtone.

In general, the oscillator tank is tuned to the crystal overtone. A better viewpoint is that the reactive elements of the tank and crystal form a coupling network between the transistor output and input, which produces the necessary phase shift. The function of the crystal is to introduce an additional reactive element capable of causing large phase shift changes for a very small frequency change. To maintain the 360° loop phase shift, changes in reactances in the circuit can be compensated for by only tiny frequency shifts, due to the extremely high crystal Q, and usually excellent temperature stability. Where extreme stability is required, the crystal should be housed in a temperature-stable environment (crystal oven).

Crystals reduce oscillator efficiency, due to losses in the crystal. These losses are represented by their series resistance. Typically, crystal series resistance is in the order of 20 to 100 Ω. Oscillator efficiency may be increased by reducing signal currents in all dissipative elements. Generally, this is done by including RF chokes (RFCs) in the transistor element leads. The values for such chokes are determined by frequency, as discused in the design examples (described in later paragraphs of this section).

8-1.3. Transistor Oscillator Design Characteristics

Many factors must be considered in the design of stable transistor oscillators. In general, all of the characteristics for vacuum tube oscillators apply to transistor oscillators. For example, the frequency determining components must be temperature stable, and mechanical movement of the individual components should not be possible.

In each of the circuits of Figs. 8-1 through 8-4, two biasing arrangements are shown. One biasing arrangement makes use of a *bypass source resistor*, the other arrangement uses a *gate resistor with a biasing diode*. Although the circuits of Figs. 8-1 through 8-4 show MOSFETs, similar design relationships exist for two-junction transistors and JFETs. The following notes and illustrations show the effect on circuit performance of the two biasing methods.

Figure 8-6 is a plot of *frequency stability* versus drain voltage using source-resistor bias. Figure 8-7 is a plot of the same conditions but for the

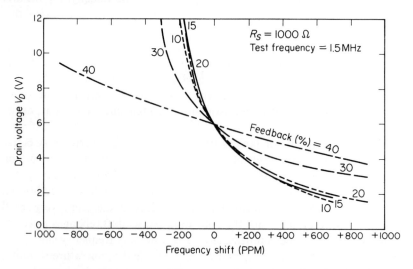

Figure 8-6 Frequency stability versus drain voltage for circuits with source-resistor bias (Courtesy RCA)

Sec. 8-1   *LC* and Crystal-Controlled Oscillators   273

circuits using a gate resistor and biasing diode. The plots clearly show the effect of various levels of feedback on oscillator stability. Note that the lowest practical feedback level is 10 per cent. With either bias method, the best feedback level is about 15 per cent. Rarely is more than 25 to 30 per cent ever required. Also note that the per cent refers to *feedback versus output voltage*.

Figure 8-8 is another plot of frequency stability showing the performance of both bias methods at a 15 per cent feedback ratio. Under normal operation, the source-resistance biasing method has a slight advantage over the gate resistor and biasing diode method. However, if the output voltage *regulation* is of any importance, the gate-resistor-biasing-diode method is superior. Figure 8-9, a plot of output voltage versus drain voltage for both biasing methods show this condition.

Figures 8-10 and 8-11 show how frequency is affected when oscillators are loaded. Note that with source-resistor bias the oscillator stops functioning at lower load levels than when bias is provided by the resistor and biasing diode.

Figure 8-12 is another plot of frequency stability showing the performance of both bias methods operating with different load at a 15 per cent feedback ratio. Note that there is very little frequency shift with a 15 per cent feedback ratio.

All of the characteristics illustrated in Figs. 8-6 through 8-12 were measured at a frequency of 1.5 MHz, using a tuning capacitor C valued at 2 pF per meter. If the operating frequency of the oscillators is increased into the VHF and UHF regions, the percentage of feedback must also be increased to compensate for the additional circuit loading. Likewise, the

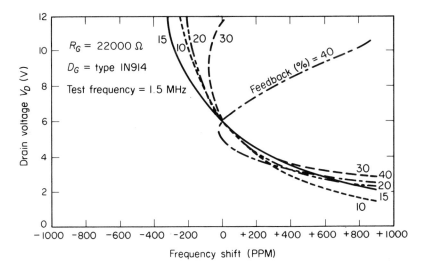

Figure 8-7 Frequency stability versus drain voltage for circuits biased with gate resistor and biasing diode (Courtesy RCA)

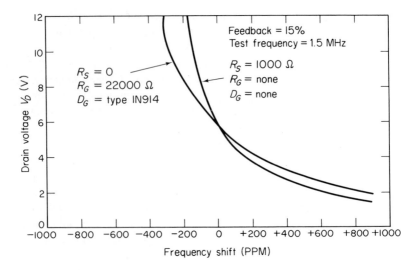

Figure 8-8 Comparison curves of 15% feedback for source-resistor and gate-resistor-biasing-diode methods (Courtesy RCA)

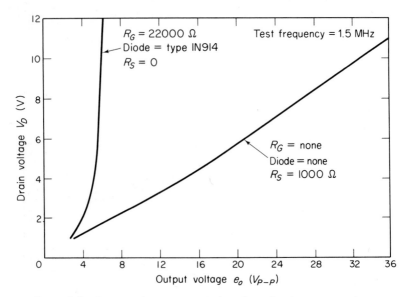

Figure 8-9 Output voltage versus drain voltage for the source-resistor and gate-resistor-biasing-diode methods (Courtesy RCA)

Sec. 8-1 　　　　　　　　　　　 LC and Crystal-Controlled Oscillators　　　275

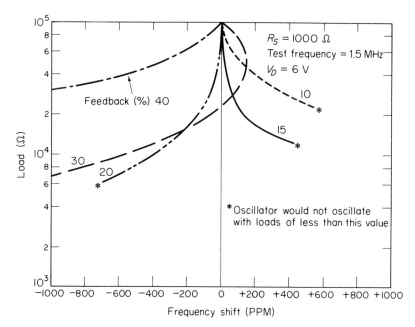

Figure 8-10 Frequency shift versus load for the source-resistor bias method (Courtesy RCA)

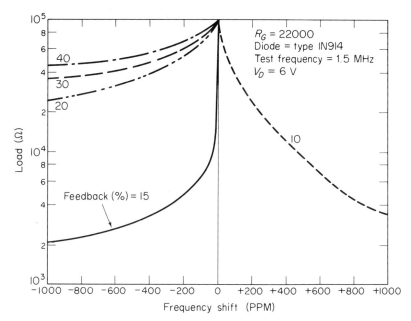

Figure 8-11 Frequency shift versus load for the gate-resistor-biasing-diode method (Courtesy RCA)

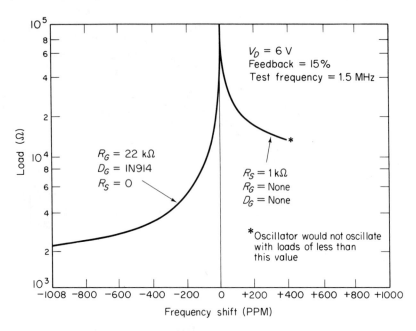

Figure 8-12 15% feedback comparison curves for both biasing methods with circuits loaded (Courtesy RCA)

percentage of feedback must be increased if the tuning circuits are made high-C through substantial increase in tuning capacitor value.

8-1.4. Practical Crystal Oscillator

Figure 8-13 is the working schematic of a crystal-controlled oscillator. This circuit is one of the many variations of the Colpitts oscillator. However, the output frequency is fixed and controlled by the crystal. The circuit can be used over a narrow range by L_1 (which is slug-tuned).

For maximum efficiency, the resonant circuit (C_1, C_2, L_1, and the transistor output capacitance) should be at the same frequency as the crystal. If reduced efficiency is acceptable, the resonant circuit can be at a higher frequency (multiple) of the crystal frequency. However, the resonant circuit should not be operated at a frequency higher than the 4th harmonic of the crystal frequency.

Bias circuit. The bias circuit components, R_1, R_2, and R_3, are selected to produce a given current flow under no-signal conditions. The bias circuit is calculated and tested on the basis of *normal operating point*, even though the circuit will never be at the operating point. A feedback signal is always present, and the transistor is always in a state of transition.

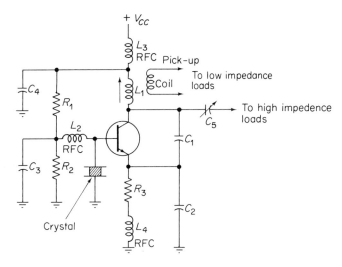

Figure 8-13 Crystal-controlled oscillator circuit design

With FETs, if temperature stability is of prime importance, the current (I_D) should be at the 0-TC point. With any transistor, the current should be set at a value to produce the required power out. With the correct bias-feedback relationship, the output power of the oscillator will be about 0.3 times the input power.

Typically, the voltage drop across L_1 and L_3 is very small, so that the collector (or drain) voltage equals the supply voltage. Thus, to find a correct value of current for a given power output and supply voltage, divide the desired output into 0.3 to find the required input power. Then divide the input power by the supply voltage to find current flow.

Feedback signal. The signal output appears at the collector or drain terminal. With the proper bias-feedback relationship, the output signal is about 90 per cent of the supply voltage. The amount of feedback is determined by the ratio of C_1 and C_2. For example, if C_1 and C_2 are of the same value, the feedback signal is one-half of the output signal. If C_2 is made about three times the value of C_1, the feedback signal is about 0.25 of the output signal voltage.

It may be necessary to change the value of C_1 in relation to C_2, in order to get a good bias-feedback relationship. For example, if C_2 is decreased in value, the feedback increases and the oscillator operates nearer the class C region. An increase in C_2, with C_1 fixed, decreases the feedback and makes the oscillator operate as class A. Keep in mind that any change in C_2 (or C_1) will also affect frequency. Thus, if the C_2/C_1 values are changed, it will probably be necessary to change the value of L_1.

As a first trial value, the amount of feedback should be equal to, or greater than, cutoff. Under normal conditions, such a level of feedback should be sufficient to overcome the fixed bias (set by R_1 and R_2) and the variable bias set by R_3. As shown in Fig. 8-6 through 8-12, feedback is generally within the limits of 10 and 40 per cent, with the best stability in the 15 to 25 per cent range.

Frequency. Frequency of the circuit is determined by the resonant frequency of L_1, C_1, and C_2, and by the crystal frequency. Note that C_1 and C_2 are in series so that the total capacitance must be found by the conventional series equation. Also note that the output capacitance of the transistor must be added to the value of C_1. At low frequencies, the output capacitance can be ignored since the value is usually quite low in relation to a typical value for C_1. At higher frequencies, the value of C_1 is lower, so the output capacitance becomes of greater importance.

For example, if the output capacitance is 5 pF at the frequency of interest, and the value of C_1 is 1000 pF, or larger, the effect of the output capacitance will be small. (Transistor output capacitance can be considered as being in parallel with C_1.) If the value of C_1 is lowered to 5 pF, the parallel output capacitance will double the value. Thus, the output capacitance must be included in the resonant-frequency calculation.

Transistor output capacitance is not always listed on datasheets. The capacitance presented by the output of a transistor (drain-to-source in the case of an FET; collector-to-emitter for a two-junction transistor) is composed of both output capacitance and reverse capacitance. However, reverse capacitance is usually small in relation to output capacitance, and can generally be ignored.

When output capacitance is not available on datasheets, it is possible to calculate an approximate value of output capacitance from output admittance (y_{os}). The imaginary part of output admittance (jb_{os} or j_{b22}) represents susceptance, which is the reciprocal of reactance. Thus, to find the reactance presented by the drain-source or collector-emitter terminals of the transistor at the datasheet frequency, divide jb_{os} into one. Then find the capacitance that will produce such reactance at the datasheet frequency using the equation

$$C = \frac{1}{6.28 F X_C}$$

where C is capacitance, F is frequency, and X_C is capacitive reactance found as the reciprocal of jb_{os}.

Of course, this method assumes that the jb_{os} reactance is capacitive, and that the capacitive remains constant at all frequencies (at least that the capacity is the same for the datasheet frequency and design frequency).

Capacitor C_1 can be made variable. However, it is generally easier to make

L_1 variable, since the tuning range of a crystal-controlled oscillator is quite small.

Typically, the value of C_2 is about three times the value of C_1 (or the combined values of C_1 and the transistor output capacitance, where applicable). Thus, the signal voltage (fed back to the source or emitter terminal) is about 0.25 of the total output signal voltage (or about 0.2 of the supply voltage, when the proper bias-feedback relationship is established).

Resonant circuit. Any number of L and C combinations could be used to produce the desired frequency. That is, the coil can be made very large or very small, with corresponding capacitor values. Often, practical limitations are placed on the resonant circuit (such as available variable inductance values).

In the absence of some specific limitations, and as a starting point for resonant-circuit values, the capacitance should be 2 pF per meter. For example, if the frequency is 30 MHz, the wavelength is 10 meters, and the capacitance should be 20 pF. Wavelength in meters is found by the equation

$$\text{Wavelength} = \frac{300}{\text{frequency (MHz)}}$$

At frequencies below about 1 to 5 MHz, the 2 pF/meter guideline may result in very large coils to produce the corresponding inductance. If so, the 2 pF/meter can be raised to 20 pF/meter.

The value of L is then selected to match a given frequency using the equation

$$L \text{ (in } \mu\text{H)} = \frac{2.53 \times 10^4}{\text{frequency } F \text{ (in MHz)}^2 \times \text{capacitance } C \text{ (in pF)}}$$

As an alternate method to find realistic values for the resonant circuit, use an inductive reactance value (for L_1) between 80 and 100 Ω at the operating frequency. This guideline is particularly useful at low frequencies (below 1 MHz).

Output circuit. Output to the following stage can be taken from L_1 by means of a pick-up coil (for low-impedance loads) or coupling capacitor (for high-impedance loads). Generally, the most convenient output scheme is to use a coupling capacitor (C_5), and make the capacitor variable. This makes it possible to couple the oscillator to a variable load (a load that changes impedance with changes in frequency).

Crystal. The crystal must, of course, be resonant at the desired operating frequency (or a sub-multiple thereof, when the circuit is used as a multiplier). Note that efficiency (power output in relation to power input) of the oscillator is reduced when the oscillator is also used as a multiplier. The crystal must be capable of withstanding the combined d-c and signal voltages at the transistor input (gate or base). As a rule, the crystal should be capable of withstanding the full supply voltage, even though the crystal will never be operated at this level.

Bypass and coupling capacitors. The values of bypass capacitors C_3 and C_4 should be such that the reactance is 5 Ω or less at the crystal operating frequency. A higher reactance (200 Ω) could be tolerated. However, due to the low crystal output, the lower reactance is preferred.

The value of C_5 should be approximately equal to the combined parallel output capacitance of the transistor and C_1. Make this the midrange value of C_5 (if C_5 is variable).

Radio-frequency chokes. The values of RFCs L_2, L_3, and L_4 should be such that the reactance is between 1000 and 3000 Ω at the operating frequency. The minimum current capacity of the chokes should be greater (by at least 10 per cent) than the maximum anticipated direct current. Note that a high reactance is desired at the operating frequency. However, at high frequencies, this can result in very large chokes that produce a large voltage drop (or are too large physically).

Crystal oscillator design example. Assume that the circuit of Fig. 8-13 is to provide an output at 50 MHz. The circuit is to be tuned by L_1. A 30-V supply is available. The crystal will not be damaged by 30 V and will operate at 50 MHz with the desired accuracy. The transistor has an output capacitance of 3 pF and will operate without damage with 30 V. The desired output power is 40 to 50 mW.

The drain or collector is operated at 30 V (ignoring the small drop across L_1 and L_3). The values of R_1, R_2, and R_3 should be chosen to provide a current that will produce 40 to 50 mW with 30 V at the drain or collector. A 45-mW output divided by 0.3 is 150 mW. Thus, the input power (and total dissipation) is 150 mW. Make certain that the transistor will permit a 150-mW dissipation at maximum anticipated temperature.

For example, assume that the transistor has a 330-mW maximum dissipation at 25°C, a maximum temperature rating of 175°C, and a 2-mW/°C derating for temperatures above 25°C. If the transistor is operated at 100°C, or 75° above the 25°C level, the transistor must be derated by 150 mW (75 × 2 mW/°C), or 330 mW − 150 mW = 180 mW. Under these conditions, the 150-mW input power dissipation is safe.

With 30 V at the drain or collector, and a desired 150-mW input power, the current must be 150 mW/30 V = 5 mA.

With a 30-V supply, the output signal should be about 24 V (30 × 0.8 = 24). Of course, this is dependent upon the bias-feedback relationship.

As a starting point, make C_2 three times the value of C_1 (plus the transistor output capacitance). With this ratio, the feedback signal will be 25 per cent of the output, or 6 V ($24 \times 0.25 = 6$). Considering the amount of fixed and variable bias supplied by the bias network, a feedback of 6 V may be large. However, the 6-V value should serve as a good starting point.

For realistic values of L and C in the resonant circuit, let $C_1 = 2$ pF/meter, or 12 pF (50 MHz = 6 meters; 300/50 = 6).

With C_1 at 12 pF, and the transistor output capacitance 3 pF, the value of C_2 is 45 pF ($12 + 3 = 15$; $15 \times 3 = 45$).

The total capacitance across L_1 is

$$\frac{1}{\frac{1}{15} + \frac{1}{45}} \approx 12 \text{pF}$$

With a value of 12 pF across L_1, the value of L_1 for resonance at 50 MHz is

$$L \text{ (in } \mu\text{H)} = \frac{2.53 \times 10^4}{(50)^2 \times 12} \approx 0.84 \ \mu\text{H}$$

For convenience, L_1 should be tunable from about 0.5 to 1.5 μH.

Keep in mind that an incorrect bias-feedback relation will result in distortion of the waveform, or low power, or both. The final test of correct operating point is a good waveform at the operating frequency, together with frequency stability at the desired output power.

The values of C_3 and C_4 should be $1/6.28 \times (50 \times 10^6) \times 5$, or 630 pF. A slightly larger value (say 1000 pF) will assure a reactance of less than 5 at the operating frequency.

The values of L_2 through L_4 should be $2000/6.28 \times (50 \times 10^6)$, or 6.3 μH nominal. Any value between about 3 and 9 μH should be satisfactory. The best test for the correct value of an RFC in an oscillator is to check for RF at the power supply side of the line, with the oscillator operating. There should be no RF, or the RF should be a fraction of 1 V (usually less than a few microvolts for a typical transistor oscillator). If RF is removed from the power supply line, the choke reactance is sufficiently high. Next, check for d-c voltage drop across the choke. The drop should be a fraction of 1 V (also in the microvolt range).

8-1.5. Practical Variable-Frequency Oscillator

Figure 8-14 is the working schematic of a variable-frequency oscillator. This circuit is also one of the many variations of the Colpitts oscillator. The circuit is chosen here for maximum stability at frequencies up to about 0.5 MHz. Oscillation is sustained by source or emitter feedback from the junction of C_1 and C_2 (as is the case for the crystal oscillator).

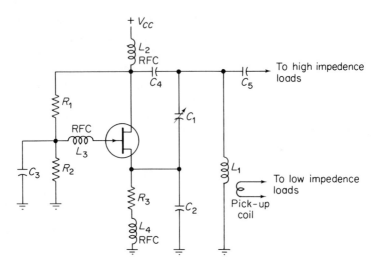

Figure 8-14 Variable-frequency oscillator circuit design

Design considerations. All of the design considerations for the variable-frequency oscillator are the same as for the crystal-controlled oscillator (Sec. 8-1.3), with the following exceptions.

Generally, C_1 is made variable to tune across a given frequency range. However, L_1 can be made variable if required.

The values of coupling and bypass capacitors C_3, C_4, and C_5 (if used) should be such that the reactance is 200 Ω at the *lowest* operating frequency (when variable capacitor C_1 is at full value). Note that capacitor C_4 and the output capacitance of the transistor may add to the C_1 capacitance. This tends to lower the resonant frequency of the L_1, C_1, C_2 circuit slightly from the calculated value. However, since C_1 is variable, there should be no problem in tuning to a desired frequency.

Variable-frequency oscillator design example. Assume that the circuit of Fig. 8-14 is to tune across a frequency range from about 10 kHz to 60 kHz. A 30-V supply is available. Thus, the transistor is operated at about 30 V (ignoring the small drop across L_2). The values of R_1, R_2, and R_3 are selected to produce the desired operating point current. Assume that the transistor has a negligible output capacitance (in relation to C_1) at the operating frequency. This is generally the case at these lower frequencies (10 to 60 kHz).

With the collector or drain at 30 V, the power input is determined by the amount of current (set by the bias network) multiplied by 30 V. Assume that the transistor is an FET and operated at an I_D of 1 mA. Under these conditions, the power input is 30 mW (30 V × 1 mA = 30 mW). Assuming a typical efficiency of 0.3, the output power is about 9 mW.

With a 30-V supply, the output signal should be about 24 V (30 × 0.8 = 24). Of course, this is dependent upon the bias-feedback relationship.

Assume that the maximum V_P (or $V_{GS(OFF)}$) is 6 V. Thus, feedback should be 6 V or greater. When C_2 is made three times the value of C_1, the feedback signal will be 6 V (24 V × 0.25 = 6 V). Considering the amount of fixed and variable bias supplied by the bias network, a feedback of 6 V may be large. However, the 6-V value should serve as a good starting point.

For realistic L and C values in the resonant circuit, the inductive reactance of L_1 should be between 80 and 100 Ω at the operating frequency. Assume a value of 100 Ω as a first trial.

With an inductive reactance of 100 and a low-frequency limit of 10 kHz, the inductance of L_1 should be $100/(6.28 \times 10 \times 10^3)$, or approximately 2 mH.

With a value of 2 mH for L_1, and a low-frequency limit of 10 kHz, the total capacitance of C_1 and C_2 (with the variable C_1 at its high limit) should be $(2.54 \times 10^4)/(10^2 \times 2000)$, or about 0.12 μF.

With a 24-V output and a 6-V feedback, the value of C_1 is 0.12(6 + 24)/24, or 0.15 μF. The value of C_2 is 0.15 × 3 = 0.45, which is rounded off to 0.5 μF.

Keep in mind that an incorrect bias-feedback relation will result in distortion of the waveform, or low power, or both. The final test of correct operating point is a good waveform at the operating frequency, together with the desired output power.

The values of C_3, C_4, and C_5 (if used) should be $1/(6.28 \times 10 \times 10^3 \times 200)$, or about 0.08-μF minimum. A slightly larger value (say 0.1 μF) will assure a reactance of less than 200 Ω at the lowest frequency.

The values of L_2 through L_4 should be $2000/(6.28 \times 10 \times 10^3)$, or about 30 mH. At the low currents involved, the 30-mH chokes should present little or no voltage drop.

8-2. RC OSCILLATORS

RC oscillators are those that use resistors and capacitors as the frequency-determining components. Figure 8-15 is the working schematic of an *RC* (resistance-capacitance) oscillator using a two-junction transistor. An FET can be used in the circuit of Fig. 8-15. However, as discussed in later paragraphs of this section, MOSFETs can be used to special advantage in certain *RC* oscillator circuits. The basic UJT relaxation oscillator described in Sec. 8-4 is also an *RC* oscillator, of sorts. However, the UJT *RC* oscillator does not produce a sine wave, as do the two-junction and FET oscillators described here.

RC oscillators are used at audio frequencies instead of the *LC* (inductance-

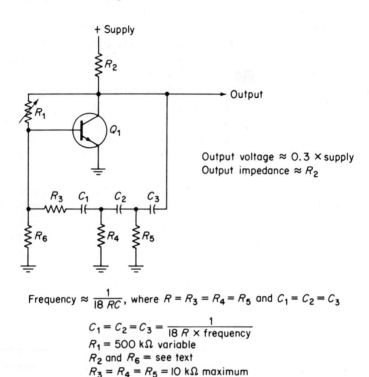

Figure 8-15 RC oscillator circuit design

capacitance) oscillators described in Sec. 8-1. *RC* oscillators avoid the use of inductances which are not practical in the audio-frequency range. Good waveforms are produced by *RC* oscillators since they generally are operated as class A.

The feedback principle is used in oscillators. In the circuit of Fig. 8-15, the collector signal is fed back through three *RC* networks to the base. The time constants of the *RC* network determine the oscillator output frequency. Each of the three *RC* networks shift the phase by about 60°, resulting in a total phase shift of 180°.

8-2.1. Design Considerations for Two-Junction Transistor RC Oscillators

In addition to the general design considerations of Sec. 8-1, the following specific points should be considered.

Power output. *RC* oscillators do not have the efficiency of *LC* oscillators because *RC* oscillators are operated class A. Also, there is considerable power loss in the *RC* network. Typically, *RC* oscillators are never more than 30 per cent efficient. Generally, power output is not a major design consideration in *RC* oscillators; voltage output is the prime factor. Typically, the voltage output is about 30 per cent of the supply voltage. Thus, the transistor-collector-voltage limits (or the available supply) set the maximum output voltage.

Bias requirements. As with any oscillator, there is no true "operating point" since there is always some signal being fed back. This presents no problems with an *RC* oscillator since it is biased class A and there is some current flowing at all times. As a guideline for bias requirements, the collector voltage should be approximately one-half the supply voltage when the collector current is at mid-point (average, or half way between minimum and maximum).

An emitter resistor is not required since there will always be some reverse bias supplied by the feedback signal (at each half-cycle). The base is set to the correct voltage by R_1. Usually, R_1 is quite large (typically 0.25 to 0.5 MΩ) since base current is in the order of 0.1 mA or less.

Output frequency. The oscillator frequency is determined by the *RC* time constants. To simplify design, the same values are used in all three *RC* networks. Thus, the output frequency is about one-third of the typical time constant frequency relationships of $1/(6.28RC)$, which results in an output frequency approximately equal to $1/(18RC)$. A more exact frequency calculation cannot be made in practical design since the transistor capacitance and resistance values are added to the *RC* network. However, the $1/(18RC)$ relationship is satisfactory for trial values.

If a variable output frequency is required, either the *C* or *R* could be made variable. However, it is common practice to make the *C* variable because three-section variable capacitors are readily available. Typically, the *RC* network resistors should not exceed about 10 kΩ. With this value, the lowest audio frequency (above 1 Hz) can be obtained with a *C* of less than 6 μF.

Transistor selection. The transistor must be capable of oscillating at the desired frequency. This presents no particular problem since most *RC* oscillators are used at frequencies below 100 kHz (generally below 20 kHz). However, the transistor must have a gain of about 60 at the operating frequency to overcome the power loss introduced by the *RC* networks.

There is a design tradeoff relationship between transistor gain and the collector resistor R_2 value. If the value of R_2 is equal to the value of the *RC* network resistors R_3, R_4, and R_5, a transistor gain of about 60 is required. If the value of R_2 is increased to about twice the value of R_3, R_4, and R_5, the required gain can be reduced to about 45. However, an increase in the value of R_2 lowers the operating-point collector voltage, thus lowering the available output-voltage swing.

If the transistor produces too much gain, resistor R_6 can be added, as shown in Fig. 8-15. As a first trial value, make R_6 the same value as R_2.

8-2.2. Design Example of Two-Junction Transistor RC Oscillators

Assume that the circuit of Fig. 8-15 is to provide a 6-V output at 60 Hz. The transistor selected is capable of oscillating over the entire audio-frequency range.

With a required output of 6 V, the supply voltage should be 6/0.3, or 20 V.

The values of R_3, R_4, and R_5 should be 10 kΩ maximum, and preferably nearer half that value. The nearest 10 per cent standard would be 4.7 kΩ.

With the RC network resistors at 4.7 kΩ, the values of the network capacitors C_1, C_2, and C_3 should be 1/(18 × 4700 × 60), or approximately 0.2 μF.

With the RC network resistors at 4.7 kΩ, the value of the collector resistor R_2 should be between 4.7 and 14 kΩ. As a first trial value, use 10 kΩ which is about halfway between the two limits.

With R_2 at 10 kΩ, and a supply of 20 V, about 1-mA collector current will be required to drop the collector voltage to half (20 V × 0.5 = 10 V; 10 V/10,000 = 1 mA).

The value of R_1 should be adjusted to provide the 1-mA current flow, as indicated when the d-c collector voltage is at 10 V. Then the output (a-c) voltage can be measured. If the transistor gain is known, the approximate value of R_1 can be calculated. For example, if the gain is 50, the value of R_1 would be 100 kΩ. This is found as follows: 1 mA (collector current)/50 = 20 μA (base current); with 20 V (supply)/20 μA base current), the value of R_1 is 100 kΩ.

As a practical matter, use a variable resistance for R_1 (arbitrarily 500 kΩ). Then adjust for the correct voltages and waveforms at the collector output.

It may be necessary to trade off R_1 and R_2 values. As a guide, if the waveform is poor, or oscillations are unstable, change the bias by changing the value of R_1. On the other hand, if the waveform and oscillations are good, but the output voltage is low or high, change the value of R_2.

If the frequency is incorrect, change the values of the capacitors and/or resistors in the RC network. This problem usually does not arise when the circuit is used as a variable frequency oscillator (where the capacitors are made variable). One exception is where a high- or low-frequency limit cannot be reached over the range of the variable capacitors. The problem does arise when the circuit is used to produce a fixed frequency. It is possible to modify the value of one RC network (say the value of R_3, R_4, or R_5) to change frequency. This may result in distortion, however. Likewise, a drastic change in only one RC network value may affect the overall phase shift and result in unstable oscillation.

8-2.3. Design Considerations for Twin-T RC Oscillators

Figure 8-16 is the working schematic of a twin-T RC oscillator using two two-junction transistors. Such circuits are used at audio frequencies where a highly stable output at a fixed frequency is desired. In addition to the general design considerations of Sec. 8-1, the following specific points should be considered.

Power output. Typically, twin-T RC oscillators are never more than about 30 per cent efficient. Likewise, the voltage output is about 30 per cent of the supply voltage. Thus, the transistor-collector-voltage limits (or the available supply) set the maximum output voltage.

Bias requirements. Both transistors should be forward-biased so that they operate class A. This will insure a good waveform. The base bias for Q_1 (an emitter follower with no voltage gain) is supplied through R_4, network

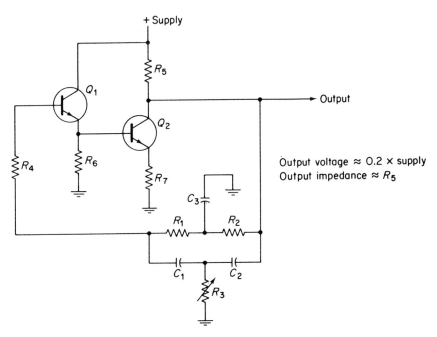

Frequency $\approx \dfrac{1}{5\,RC}$, where $R = R_1 = R_2$, $R_3 = 0.5 R_1$ and $C = C_1 = C_2$, $C_3 = 2C$

$R_4 \approx R_1 + R_2$ $\qquad R_5 \approx 60\,R_7$
$R_6 \approx 10\,R_7$ $\qquad R_7 \approx 50 - 100\,\Omega$
$R_1 = R_2 = 1000\,R_7$ $\quad C_1 = C_2 = \dfrac{1}{5\,R_1 \times \text{frequency}}$
$R_3 = 0.5\,R_1$ $\qquad C_3 = 2C_1$

Figure 8-16 Twin-T RC oscillator circuit design

resistors R_1 and R_2, and collector resistor R_5. As a rule, R_4 should equal the total series resistance of R_1 and R_2. The collector of Q_1 is connected directly to the source, while the Q_1 emitter is returned through the base resistance of Q_2.

Base resistor R_6 should be approximately ten times the value of R_1. Transistor Q_2 should be biased similar to class A. However, considerable voltage gain is required to overcome the network loss. Thus, R_5 (collector) should be approximately 60 times the value of R_7 (emitter) for maximum voltage gain. For a realistic R_5 value, the value of R_7 should be less than 100 Ω (typically 50 Ω or the nearest standard).

Any number of RC combinations could be used to produce the desired operating frequency. However, since R_1 and R_2 also form part of the bias network, design of the RC network should start with these resistors. As first trial values, R_1 and R_2 should be 1000 times the value of R_7.

Output frequency. The oscillator frequency is determined by the RC time constant of the twin-T network. This network is essentially a filter that has a sharp null, or balance at the resonant frequency, as determined by $F = 1/(6.28RC)$. By decreasing the value of shunt resistor R_3 slightly from the balance point, the output of the twin-T network is a small, in-phase signal that is rapidly changing in phase as the balance frequency. When the network is adjusted off-balance (the normal oscillating condition) by R_3, the approximate frequency is found by $F \approx 1/(5RC)$.

Transistor selection. The transistors must be capable of oscillating at the desired frequency. This presents no particular problem since most RC oscillators are used at frequencies below 100 kHz (generally below 20 kHz). Both Q_1 and Q_2 can be the same transistor type, if convenient. However, Q_2 must have a gain of about 100 at the operating frequency to overcome the power loss.

8-2.4. Design Example of Twin-T RC Oscillator

Assume that the circuit of Fig. 8-16 is to provide an output of 60 Hz. The transistors are capable of oscillating over the entire audio range with the full supply voltage applied.

The key design value is the resistance of R_7. If the remaining resistor values are to be kept within reason, the value of R_7 should be less than 100 Ω. This will provide just enough reverse bias to stabilize both Q_1 and Q_2. A much larger value will usually provide so much reverse bias that the circuit cannot oscillate.

With R_7 at 51 Ω (the nearest standard value at approximately half of 100 Ω), the values of the remaining resistors are: R_1 and R_2 51 kΩ, R_3 25 kΩ (maximum), R_4 100 kΩ, R_5 3 to 3.3 kΩ, and R_6 510 Ω.

Sec. 8-2   RC Oscillators   289

With R_1 and R_2 at 51 Ω

$$C_1 = C_2 = \frac{1}{5 \times (51 \times 10^3) \times 60} = 0.065 \ \mu F$$

With C_1 and C_2 at 0.065 μF, the value of C_3 should be 0.065 × 2, or 0.13 μF.

The critical element of this circuit is the adjustment of R_3. Resistor R_3 should always be made variable during the breadboard stage of design, even if the remaining resistance values are fixed. The value of R_3 should be increased until the circuit stops oscillating; then reduce the value of R_3 slightly (detuning the RC filter) to allow an increase in the 60-Hz signal to sustain oscillation.

If it is necessary to detune the filter (change the value of R_3 from the non-oscillating point) by a large amount to get stable oscillations, the overall circuit gain is too low. This condition can be corrected by decreasing the value of R_4, increasing the value of R_5, or changing the transistor for higher gain.

With the correct value of R_3 chosen, the oscillations should remain stable despite changes in supply voltage (within reasonable limits). Using the values described, it should be possible to vary the supply voltage from about 12 to 28 V, and still maintain stable oscillation. Of course, the output voltage will vary with changes in supply. Generally, the output voltage will be about 0.2 times the supply voltage. However, this ratio does not always remain true. A higher percentage (possibly 0.25 or 0.3) can often be obtained by increasing the supply, all other factors being equal.

8-2.5. Design Considerations for MOSFET RC Oscillator

Figure 8-17 is the working schematic of a MOSFET RC phase-shift oscillator. MOSFETs are well suited to RC circuits since no

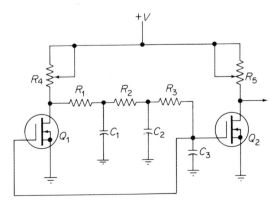

Figure 8-17 MOSFET RC phase-shift oscillator

coupling capacitors are needed between stages (the MOSFET gate acts as a capacitor).

The feedback principle is also used in MOSFET *RC* oscillators. In the circuit of Fig. 8-17, the output (drain) of Q_1 is fed through three *RC* networks back to the gate of Q_1. Each network shifts the phase about 60°, resulting in an approximate 180° shift between drain and gate. Since the drain is normally shifted 180° from the gate, the *RC* shift of 180° brings the feedback to 360°, or back in-phase to produce oscillation. MOSFET Q_2 is used as an output amplifier.

Bias requirements. Since both MOSFETs are operated at zero gate voltage, the *Q*-point drain voltage is set by I_{DSS} and the values of R_4 and R_5. Both R_4 and R_5 are made variable. Resistor R_4 is adjusted to produce oscillations with good waveforms. Resistor R_5 is adjusted to produce the desired output swing. Typically, both Q_1 and Q_2 should be operated at one-half the supply voltage. For example, if I_{DSS} is 1 mA, and the supply is 30 V, both R_4 and R_5 should be 15 kΩ, thus dropping both drains to about 15 V.

Output frequency. The oscillator frequency is determined by the *RC* time constant. To simplify design, the same values can be used in all three *RC* networks. However, such an arrangement will create a problem of power loss. Each of the *RC* networks functions as a low-pass filter. If the same values are used in all three sections, the signal loss will be about 15 dB through the networks.

Such a loss, combined with the normal loss, could be sufficient to prevent oscillation if the voltage gain of Q_1 is low. The loop gain of any oscillator must be at least 1 (or slightly more for practical design). If the gain of Q_1 is 10 and the loss is anything greater than about 8 to 8.5, the circuit will not oscillate.

The *RC* network loss problem can be minimized by making the impedance of the succeeding network *greater* than that of the prior network. That is, there should be an impedance step-up of a multiple of three as the signal passes from the drain of Q_1 to the gate of Q_2. For example, R_2 should be three times that of R_1; R_3 should be three times that of R_2. Thus, each *RC* network places very little load on the previous section and keeps loss at a minimum.

There should also be an impedance step-up between the output of Q_1 (set by R_4) and the first *RC* network. As a first trial, R_1 should be at least three times the value of R_4.

The values of C_1, C_2, and C_3 must be selected to produce the desired operating frequency. The output frequency is about equal to $1/(3RC)$. A more exact frequency calculation cannot be made in practical design since MOSFET capacitance and resistance values must be added to the *RC* networks. However, the $1/(3RC)$ relationship is satisfactory for trial values.

8-2.6. Design Example of MOSFET RC Oscillators

Assume that the circuit of Fig. 8-17 is to provide an output at 3.7 kHz. The power supply is 30 V. The output signal is to be the maximum possible without distortion. The MOSFETs have a zero gate voltage I_D of 1 mA.

For maximum output voltage swing, the drains of both Q_1 and Q_2 should be at one-half the supply or 15 V.

With 1 mA of I_D flowing, the drops across R_4 and R_5 should be 15 V. Thus, R_4 and R_5 should be 15 kΩ (15 V/0.001 = 15 kΩ).

With R_4 at 15 kΩ, R_1 should be at least 45 kΩ. With R_1 at 45 kΩ and a 3.7 kHz operating frequency, C_1 should be approximately 0.002 μF; 1/(3 × 45 kΩ × 3.7 kHz) = 0.002 μF.

With R_1 at 45 kΩ, R_2 should be 135 kΩ (to get the impedance step-up). With R_2 at 135 kΩ and a 3.7-kHz operating frequency, C_2 should be approximately 7000 pF; 1/(3 × 135 kΩ × 3.7 kHz) = 7000 pF.

With R_2 at 135 kΩ, R_3 should be 405 kΩ. With R_3 at 405 kΩ and a 3.7 kHz operating frequency, C_3 should be approximately 2200 pF; 1/(3 × 405 kΩ × 3.7 kHz) = 2200 pF.

In practice, after the values have been selected and the components assembled, the gate of Q_2 is monitored on an oscilloscope, and R_4 is adjusted for maximum signal swing without distortion; R_5 is then adjusted for maximum output swing, without distortion, at the drain of Q_2.

8-3. BLOCKING OSCILLATORS

Both FETs and two-junction transistors can be used as blocking oscillators. However, the two-junction transistor is generally used for the blocking oscillator. The basic UJT oscillator described in Sec. 8-4 operates on the *relaxation oscillator* principle, which is similar to the blocking oscillator operation covered here.

Fig. 8-18 is the working schematic of a blocking oscillator using a two-junction transistor. The same basic principles can be applied to FETs (particularly JFETs) if desired.

Note that two versions of the circuit are shown, one with a tapped transformer. The blocking oscillator is one of the simplest solid-state oscillators and operates on the relaxation principle. The transistor is initially forward-biased to conduct through the transformer primary winding. This causes a signal to be fed back to the base through C_1. The base is driven very hard in the forward direction, so that C_1 is rapidly charged through the forward-biased emitter-base junction.

The output pulse at the T_1 secondary is generated by the rapid turn-on of

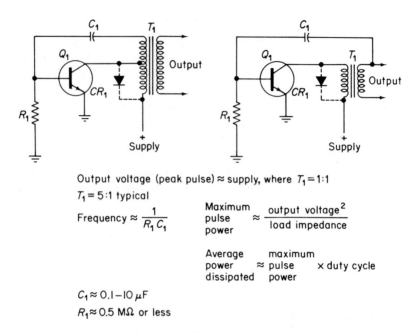

Output voltage (peak pulse) ≈ supply, where $T_1 = 1:1$
$T_1 = 5:1$ typical

Frequency ≈ $\frac{1}{R_1 C_1}$

Maximum pulse power ≈ $\frac{\text{output voltage}^2}{\text{load impedance}}$

Average power dissipated ≈ maximum pulse power × duty cycle

$C_1 ≈ 0.1 - 10 \, \mu F$
$R_1 ≈ 0.5 \, M\Omega$ or less

Figure 8-18 Free-running blocking oscillator circuit design

the collector current, and the steepness of the pulse wavefront is limited only by the leakage inductance of T_1. The top of the pulse is flattened by collector current saturation. When saturation is reached, the feedback signal drops and is no longer able to forward bias the transistor. Capacitor C_1, in discharging, reverse biases Q_1 and discharges slowly through R_1. The cycle is re-started by forward bias.

The time constant $R_1 C_1$ determines the off-time between pulses and thus determines the oscillator frequency.

The output of a blocking oscillator is similar to that of Fig. 8-19. Blocking oscillators are not suited for radio frequencies or an application where a sine wave (or anything approaching a sine wave) is needed. However, blocking oscillators are excellent sources of the steep wavefront pulses required in switching applications. A pulse rise time of 0.1 μs is not uncommon for blocking oscillator outputs.

Blocking oscillators can be used as driven oscillators where a single output pulse is produced in response to a trigger pulse. This requires that the transistor emitter base be reverse-biased, as shown in Fig. 8-20. With this arrangement, the transistor is initially cut off so that a trigger signal moves the operating point into the active region just long enough to start a pulse cycle.

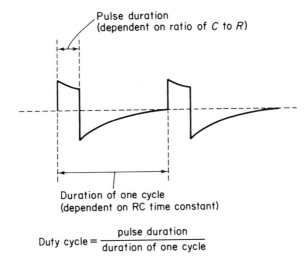

Figure 8-19 Typical blocking oscillator output waveform

One of the major advantages of a blocking oscillator is that very little current is drawn between pulses. Thus, high pulse currents can be drawn from the output for short durations without exceeding the power capability of the transistor.

8-3.1. Design Considerations for Blocking Oscillators

In addition to the general design considerations of Sec. 8-1, the following specific points should be considered.

Voltage output. If the primary and secondary windings of T_1 have the same number of turns (a 1-to-1 turns ratio), the output voltage (pulse peak) will be approximately equal to the supply voltage. Generally, the secondary winding of T_1 has fewer turns than the primary to provide a voltage step down.

When the untapped transformer version of the circuit (Fig. 8-18) is used (where C_1 is connected to the secondary), the full output voltage will be applied to the base during the brief pulse peak. In some cases, this could result in damage to the transistor. As a general rule, a 5:1 turns ratio is used so that the output voltage is about 1/5 of the supply voltage.

Transformer selection. Any transformer can be used with a blocking oscillator, provided the primary can withstand the voltage and current, and the secondary is at the desired output impedance. There is a momentary voltage surge across the primary winding equal to approximately twice the supply voltage. Often, special purpose transformers are designed for use with blocking oscillators. Such transformers are supplied with design data for the blocking oscillator circuit. This information should be followed when available.

Transistor selection. The transistor must be capable of oscillating at the desired frequency and must be capable of withstanding the full supply voltage continuously. The momentary buildup across the transformer primary is also applied to the transistor. Thus, the collector may be at about twice the supply voltage during the pulse peak. This condition can be corrected by the addition of diode CR_1 across the primary, as shown in dotted form on Fig. 8-18. Diode CR_1 must be capable of withstanding twice the supply voltage without breakdown. Since any diode will have some leakage, as well as some foward voltage drop, the addition of CR_1 to the circuit can result in a drop of the output voltage.

Current through the transistor and the resultant power dissipation is difficult to calculate in a blocking oscillator. As a rough approximation, divide the square of the anticipated output voltage (across the transformer secondary) by the anticipated load impedance to find the maximum pulse power. The average power dissipated by the transistor will then depend upon the duration of the pulse in relation to the spacing between pulses.

For example, assume that the output load impedance is 50 Ω, the output voltage is 10 V, and the pulse duration is 1 mS with a frequency of 100 Hz. The peak power dissipation is 10 $V^2/50$, or 2 W. With 1-mS pulses spaced by 9 mS, the transistor is on 0.1 of the time. Thus, the average power dissipated by the transistor is 2 W $\times$ 0.1, or 0.2 W.

Operating frequency. The operating frequency is determined by the time constant of R_1C_1 and is approximately equal to the reciprocal of the time constant. The exact frequency is difficult to calculate because both the transistor and transformer characteristics can affect the charge and discharge function. Also, the supply voltage can have some effect on frequency. However, blocking oscillators are fairly stable with respect to power supply variations.

Various combinations of R and C can be used to produce a given time constant (and thus a given frequency). However, the following rules should be applied.

The value of C_1 should be between 0.1 and 10 μF, with the value of R_1 less than 0.5 MΩ, for a free-running blocking oscillator in the audio-frequency range.

When a larger value of C is used (with an R of corresponding low value) to produce a given RC time constant, the pulse duration will be longer in relation to the complete cycle. That is, the on-time (or duty cycle) will be longer. This increases the average power dissipation as well as the average power output.

In practical design, the circuit should be tested in breadboard form using the desired transformer, transistor, and supply voltage. The value of C should be fixed (at 10 μF for a first trial), and R_1 should be made variable (say a 0.5 MΩ potentiometer). Adjust R_1 to the approximate value required to produce the operating frequency. Apply power and observe the waveform for amplitude, pulse duration, and frequency.

If the frequency is incorrect, adjust R_1 until the desired frequency is obtained.

If the pulse duration is too long (with amplitude and frequency correct) decrease the value of C, and increase the value of R by a corresponding amount.

If there is a sharp spike (or overshoot) on either edge of the pulse, connect diode CR_1 across the transformer primary.

Bias requirements. The free-running version of the blocking oscillator (Fig. 8-18) is initially forward-biased through R_1. The amount of bias is not critical. Once the circuit begins to oscillate, the emitter-base junction is driven into full forward bias and full reverse bias by the charge and discharge of C_1.

The driven oscillator version of the blocking oscillator (Fig. 8-20) requires a fixed reverse bias on the emitter-base junction. This bias must be less than

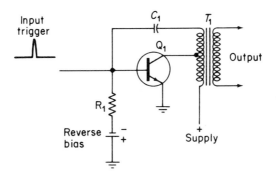

Output voltage (peak pulse) $\approx$ supply, where $T_1 = 1:1$

Frequency $\approx \dfrac{1}{R_1 C_1}$ $\quad$ Maximum pulse power $\approx \dfrac{\text{output voltage}^2}{\text{load impedance}}$

$C_1 \approx 0.1 - 10 \, \mu$F

$R_1 \approx 0.5$ MΩ or less $\approx$ input impedance of oscillator

Average power dissipated $\approx$ maximum pulse power $\times$ duty cycle

Reverse bias $\approx$ peak trigger input $\times$ 0.5

$C_1 \approx \dfrac{1}{\text{freq.} \times R}$

Figure 8-20 Triggered or driven blocking oscillator circuit design

the available trigger source. As a first trial, make the reverse bias voltage one-half the trigger voltage.

8-3.2. Design Example of a Free-Running Blocking Oscillator

Assume that the circuit of Fig. 8-18 is to provide a pulse output of 4 V across a 50-Ω load using a transformer with an untapped primary. The operating frequency is 1 kHz.

Although other turns ratios could be used, a 5:1 ratio is typical. That is, the transformer secondary should have a 50-Ω impedance to match the load, and the primary should have five times as many turns as the secondary. The primary impedance would be 1250 Ω; primary impedance/secondary impedance = (primary turns/secondary turns)^2, or $(5/1)^2 = 25$; $25 \times 50 = 1250$.

With a required 4 V at the secondary and a turns ratio of 5:1, the primary voltage is 20 V. Thus, the supply voltage (and the collector voltage) is 20 V.

With 4 V and 50 Ω at the secondary, the secondary current is 80 mA, and the maximum power dissipation is 320 mW.

With a 5:1 turns ratio and a secondary current of 80 mA, the primary current (and collector current) is 16 mA.

If both the transformer and transistor are selected on the basis of the maximum voltage, current, and power dissipation calculations, there will be ample margin for safety. In practice, the transistor need not be capable of dissipating the full 320 mW. About half that value will still provide considerable safety unless the pulse duration approached half the full cycle between pulses. Usually, blocking oscillators are operated at 0.1 to 0.2 duty cycles. Thus, the true dissipation would probably be more on the order of 32 to 64 mW.

The value of C_1 should be kept between the limits of 0.1 to 10 μF. Since the desired operating frequency is 1 kHz, the lower value (0.1 μF) should be used as the first trial. With a value of 0.1 μF for C_1, the value of R_1 should be: $1/(1000 \times 0.1 \times 10^{-6})$, or 10 k$\Omega$.

As discussed, R_1 must be made variable in the breadboard stage and then adjusted for the desired frequency. With the frequency established, check for proper waveform, pulse duration, and amplitude. Change the values of C and R if necessary.

8-3.3. Design Example of a Driven Blocking Oscillator

Assume that the circuit of Fig. 8-20 is to provide a pulse output of 10 V across a 50-Ω load using a transformer with an untapped primary. The circuit is to be driven at a rate of 3 kHz by a trigger pulse of 3 V. The trigger source impedance is 15 kΩ. The available supply is 20 V.

With a 20-V supply and a 10-V output required, the transformer turns ratio must be 2-to-1. That is, the transformer secondary should have a 50-Ω impedance to match the load, and the primary should have twice as many turns as the secondary. The primary impedance is approximately 200 Ω.

With 10 V and 50 Ω at the secondary, the secondary current is 200 mA, and the maximum power dissipation is 2 W.

With a 2-to-1 turns ratio and a secondary current of 200 mA, the primary current (and collector current) is 100 mA.

As in the case of the free-running oscillator, if both the transformer and transistor are selected on the basis of the maximum voltage, current, and power dissipation calculations, there will be an ample margin for safety. In practice, the transistor need not be capable of dissipating the full 2 W. About half that value will still provide considerable safety.

The value of R_1 should be approximately 15 kΩ to match the trigger source impedance. With a value of 15 kΩ for R_1 and a trigger frequency of 3 kHz, the value of C_1 should be $1/(3000 \times 15,000)$ or 0.022 μF. Note that this is lower than the nominal 0.1 μF low limit for free-running oscillators operating the audio range. However, in a driven oscillator the tradeoff should be such that the RC values match the driving frequency.

With a 3-V trigger pulse, the reverse bias should be 3×0.5, or 1.5 V.

The reverse bias should be made variable in the experimental stage and then adjusted for the correct operating point. A high reverse bias may prevent the circuit from being triggered. A low reverse bias may cause the circuit to trigger at the wrong point on the trigger signal, or to be triggered by the undesired signals mixed with the trigger source.

With the reverse bias established at the correct point, check the output waveform. Adjust the values of C and R if necessary.

8-4. BASIC UNIJUNCTION RELAXATION OSCILLATOR

The relaxation oscillator is the basic building block in most unijunction transistor timer and oscillator circuits. Figure 8-21 shows the basic circuit along with some typical waveforms. In this section we shall discuss the design considerations for selection of the basic component parts in the UJT circuit. For a more detailed discussion of UJT oscillator and timer circuits, the reader's attention is invited to the author's *Handbook for Transistors* (Prentice-Hall, Inc., Englewood Cliffs, N.J., 1976).

8-4.1. Design Considerations

The period of oscillation (and thus the operating frequency) of the UJT circuit is determined primarily by the values of R_E and C_E, as shown by the equations of Fig. 8-21. It is most practical to start design by selecting

298   Oscillator Circuit Design   Chap. 8

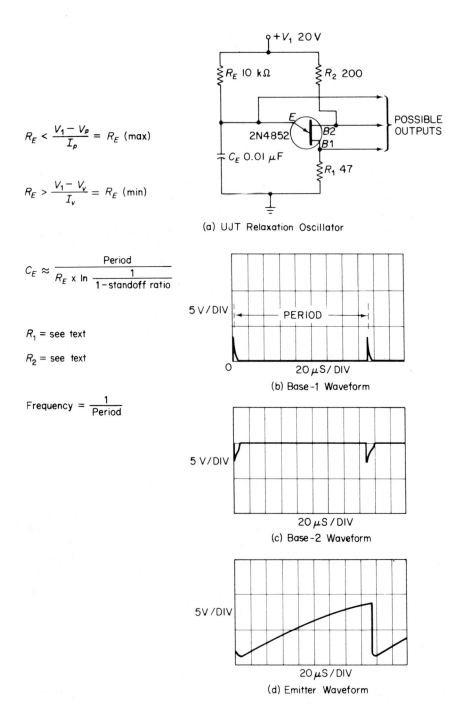

$$R_E < \frac{V_1 - V_p}{I_p} = R_E \text{ (max)}$$

$$R_E > \frac{V_1 - V_v}{I_v} = R_E \text{ (min)}$$

$$C_E \approx \frac{\text{Period}}{R_E \times \ln \frac{1}{1 - \text{standoff ratio}}}$$

R_1 = see text

R_2 = see text

$$\text{Frequency} = \frac{1}{\text{Period}}$$

Figure 8-21   Basic UJT relaxation oscillator (Courtesy Motorola)

Sec. 8-4   Basic Unijunction Relaxation Oscillator   299

a trial value for R_E rather than for C_E, because R_E must meet certain conditions for the oscillator to operate. If R_E is too large, the UJT will never fire; if R_E is too small, the UJT will not turn OFF.

These conditions can best be explained by means of the emitter characteristic curve of Fig. 8-22. (This curve is not drawn to scale in order to have more detail.)

Selecting emitter resistor R_E. The emitter capacitor C_E will charge until the emitter voltage is equal to V_P. At this point on the characteristic curve, peak-point emitter current i_p will be flowing, and in order to fire the UJT, the value of R_E must be small enough to allow a current somewhat larger than i_p to flow. R_E must, therefore, meet the following requirement

$$R_E < \frac{V_1 - V_P}{I_P} = R_{E(\max)} \qquad (8\text{-}1)$$

where V_1 is the applied bias voltage.

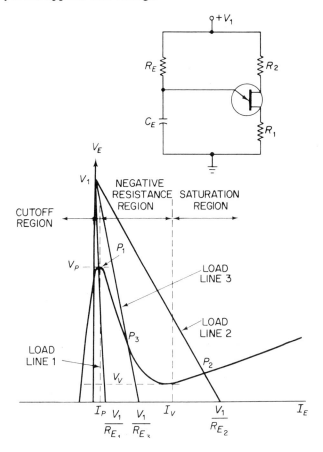

Figure 8-22   UJT emitter characteristic load lines (Courtesy Motorola)

Referring to Fig. 8-22, a load line intersecting the characteristic curve in the *cutoff region*, as illustrated by load line 1, would keep the UJT from ever firing. Therefore, R_E must be small enough to keep the UJT in the *negative resistance region*.

By keeping R_E smaller than $R_{E(max)}$, the UJT will turn ON, and C_E will discharge through the emitter. Hower, if R_E is too small, and an emitter current larger than the valley current I_V flows, the UJT will not turn OFF. Under these conditions, the UJT will operate in the *saturation region*, as illustrated by load line 2 in Fig. 8-22. The minimum R_E that can be used in order to assure oscillation is set by the following condition

$$R_E > \frac{V_1 - V_V}{I_V} = R_{E(min)} \qquad (8\text{-}2)$$

where V_1 is the applied bias voltage.

An emitter resistance R_E is selected to meet the requirements in Equations 8-1 and 8-2 will result in a load line that intersects the characteristic curve somewhat in the *negative resistance region*. This is illustrated by load line 3 in Fig. 8-22.

In a practical UJT oscillator, the emitter voltage variations in the neighborhood of the valley point are small. Thus, in order to assure turn-off, the value of R_E should be *two to three times larger than* $R_{E(min)}$, but less than $R_{E(max)}$.

Selecting emitter capacitor C_E. The value of capacitor C_E is determined by the desired period (or frequency) operation. The equation for C_E shown on Fig. 8-21 is *very approximate* since it does not take into account the ON and OFF times of the UJT, or other characteristics of the circuit such as source voltage (V_1), emitter to Base 1 voltage drop (V_D), valley voltage (V_V) standoff ratio, or interbase voltage (V_{B2B1} or, simply V_{BB}). However, the period equation in Fig. 8-21 for C_E is accurate enough for a first trial value.

Selecting Base 1 resistor R_1. The primary function of R_1 (in Fig. 8-21 or 8-22) is to provide an output load for the oscillator. In some applications, R_1 is included to provide a path for the interbase current. Such a path is necessary in some circuits to prevent current from flowing through a device being driven by the UJT oscillator (such as an SCR or other thyristor).

Typically, R_1 is less than 100 Ω, but could be as high as 2 or 3 kΩ in some applications.

When R_1 is selected on the basis of a current path for an external device, a maximum voltage drop is usually specified. For example, if the UJT oscillator is to trigger an SCR (from an output pulse across R_1), and the fixed SCR voltage must not exceed 50 mV. Assuming a 2.5-mA interbase current, and the 50-mV maximum voltage drop, the value of R_1 would be 20 Ω.

When R_1 is selected on the basis of output voltage, which is usually the case, a minimum output voltage is usually specified. For example, if the UJT oscillator is to trigger an SCR or other device, and the SCR requires a minimum of 3 V for the trigger, the peak drop across R_1 must be 3 V.

Figure 8-23 is provided to show the peak voltage across R_1, as a function of C_E, for various (typical) values of R_1. These are *minimum peak values*. In practice, the peak output across R_1 will usually be 25 to 40 per cent higher. The values shown are for a supply voltage of 20 V. The peak amplitude at other values of supply voltage V_1 can be obtained by multiplying the values from Fig. 8-23 by the factor shown.

For example, if the supply voltage is 25 V, the peak output for an R_1 of 10 Ω and a C_E of 1 μF is about 4.2 V. This is found as follows: the curve of Fig. 8-23 shows a minimum output of 3 V for an R_1 of 10 Ω, and a C_E of 1 μF. Using the correction factor for 25 V of approximately 1.4, $(25 - 6)/14$, the approximate minimum output voltage is $1.4 \times 3 \text{ V} = 4.2 \text{ V}$.

Selecting Base 2 resistor R_2. The primary function of R_2 (in Fig. 8-21 or 8-22) is to provide temperature compensation. Practically all UJT character-

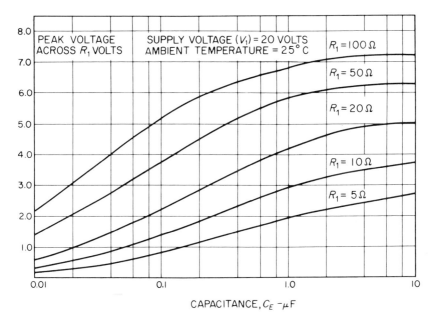

Figure 8-23 Peak output voltage across R_1 in UJT trigger circuit (minimum values) (Courtesy General Electric)

istics are temperature dependent, some more than others. The interbase resistance and emitter reverse current increase, whereas the peak and valley voltages (and currents), the intrinsic stand-off ratio, and the junction diode drop decrease with increasing temperature.

Generally, oscillator frequency is the main factor affected by the temperature variations. However, output voltage can also be affected and is of some concern in many applications.

If R_2 is properly selected, the peak point voltage V_P can be made to vary less than 1 per cent over a 50°C temperature variation. The value of R_2 can be selected by using either the equation, or the curves of Fig. 8-24.

The curves of Fig. 8-24 show frequency variation as a function of temperature for a typical UJT. Temperature curves for several values of R_2 ranging from 250 Ω to 3 kΩ are shown, and an R_2 of approximately 1.5 kΩ can be seen to compensate very well from -5°C to $+85$°C. A smaller resistance should be used for operation below -5°C.

The equation of Fig. 8-24 can be used for a somewhat more accurate "optimum" value of R_2. The equation takes into account source voltage, interbase resistance, and standoff ratio. Keep in mind that both the curves and equation of Fig. 8-24 are for trial values. Also keep in mind that an increase in R_2 value will decrease interbase voltage. As discussed, this increases frequency, but decreases output voltage (all other factors being equal).

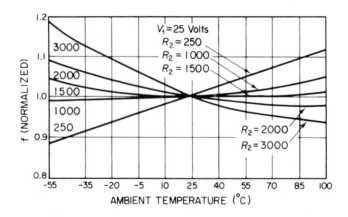

$$R_2 \approx 0.015 \times V_1 \times R_{BB} \times \text{STANDOFF RATIO}$$

Figure 8-24 Frequency versus temperature for a UJT relaxation oscillator where frequency is normalized to 25°C, and R_2 is a variable parameter (Courtesy Motorola)

INDEX

A

A-C response 160
Active filter 1, 17
Admittance
  input 34
  measurements 37
  output 36
  parameters 179
AF amplifier 107
Algebra, vector 32
Amplifier
  AF 107
  coupling 112
  design classifications 116
  efficiency 150
  line operated 166
  multistage 131
  neutralized 200
  power supplies 164
  RF 170
Analog switch 216
Analog switch, FET 223
Area light sources, photo-
    transistor 88
Asymmetrical attenuator (pad) 2
Attenuator 1, 20

Audio amplifier 157, 161
AVC-AGC 183

B

Band-elimination filter 11
Band-pass filter 4, 13
Band-pass filter (active) 20
Base injection 177
Bias
  multivibrator 68
  oscillators 267, 276
  resistance networks (RF) 173
Blocking oscillator 291
Breakdown protection 167
Bridged pads 24
Bridge-type inverter 244
Bypass
  capacitors 109, 173
  emitter 125
  source 128

C

Candle power 96
Capacitance frequency limitations 109
Capacitor coupling 113
Cascade amplifiers 131

Index

Chopper 207
  drift 214
  dual-gate 231
  FET 215, 221
  high-frequency 229
  high-input voltage 227
  JFET 226
  low-input voltage 230
  MOSFET 223, 228
Circuit design 26
Class A operation 117
Class AB operation 120
Class B operation 118
Class C operation 121
Clipping 132
Closed-loop gain 159
Coils, RF 30
Color temperature 92
Colpitts oscillator 265
Common-emitter Darlington 138
Commutator, FET switch 235
Complementary amplifier 134
Complementary symmetry circuit 156
Conjugate 38
Conjugate match 40
Constant-K filter 7
Conversion gain 175, 182
Converter
  one-transformer 255
  RF 174
  transistor 235
  two-transformer 257
Coupling
  amplifier 112
  capacitors 110, 156
  effects of 115
Crossover distortion 149, 159, 163
Crystal-controlled oscillator 265
Crystal oscillator 276, 280
Cutoff, low-frequency 133

D

Darlington compound 136, 138
Datasheet graphs or RF 194
D-C amplifier 134

D-C amplifier (FET) 139
Decoupling capacitors 109
Direct coupled amplifier 134
Direct coupling 113
Distortion 132
Driven inverter 241
Drift in choppers 214
Dual-gate chopper 231

E

Efficiency, amplifier 150
Emitter bypass 125
Emitter follower 137
Emitter injection 177
Emitter resistance 172

F

Feedback 123, 127, 133
Feedback oscillator 277
Feedback problems 50
FET amplifier 127, 140
FET (field effect transistor)
  amplifier 127, 140
  analog switch 223
  chopper 215, 221
  switch 215
  with two-junction transistor 144
Fiber optics, phototransistor 90
Filter 1
Filter capacitance, effects of 165
Filter RC 111
FM tuner 203
Forward transadmittance 34
Four-transistor amplifier 161
Frequency, AF amplifier 107
Frequency calculations for filter 16
Frequency multivibrator 68
Frequency oscillator 278
Frequency response 115
Frequency stability 272

G

Gain, conversion 175, 182
Gain versus stability 125

Gallium arsenide (GaAs)   96
General power gain   41

H

Harmonics in oscillators   269
Hartley oscillator   266
High-current multivibrator   66
High-cut filter   3
High-cut filter (active)   18
High-frequency design, phototransistor   99
High-pass filter   2, 9, 15, 111
High-pass filter (active)   19
H-pad   22
H-pad (bridged)   24
Hybrid circuits   135, 144

I

Illumination sources, phototransistor   85
Impedance
  coupling   114
  match   50
  matching, RF   191
  stray   112
Inductance frequency limits   112
Inductive coupling   114
Inductor inverter   243
Injection, mixer   177
Input admittance   34
Input transformer   154
Inverter
  bridge type   244
  common-base and common-collector   238
  component selecting   250
  design   250, 254
  driven   241
  resistive-coupled   242
  saturable base inductor   243
  series   244
  single transformer   241
  specifications   246
  speed-up circuits   248
  starting circuit   247
  transformer   251
  transistor   235
  two-transistor   239
Irradiance, effective   92
Irradiance, phototransistor   89

J

JFET chopper   226

L

LC filters   5
LC oscillators   265
LED (light emitting diode)   96
LED, phototransistor logic   105
Lens systems, phototransistor   89
L filter   7
Light sources, phototransistor   87
Linearity of sawtooth oscillators   64
Line operation amplifier   166
Linvill C factor   39
Local feedback   123
Local oscillator injection   175
Logic circuits, phototransistor   105
Loop feedback   124
Loudspeaker grounding   161
Low-current multivibrator   67
Low-cut filter   2, 111
Low-cut filter (active)   19
Low-frequency cutoff   133
Low-frequency design, phototransistor   97
Low-pass filter   3, 14
Low-pass filter (active)   18
Low-pass filter (LC)   7
Low power amplifier   156
L-pad   20

M

Maximum available gain   43
Maximum usable gain   43
m-derived filter   14
Miller effect   50
Mixer circuit   179
Mixer RF   174

MOSFET amplifier   200
MOSFET chopper   223, 228
MOSFET oscillator   283, 289
Multiplier, RF   184
Multistage amplifiers   131, 146
Multistage Darlington   139
Multistage RC filters   5
Multivibrators   66

N

Network characteristics   192
Networks, RF   180
Neutralization   40
Neutralized amplifier   200
Neutralized solution   43
Non-blocking amplifier   145

O

Offset current   209
Offset voltage   209
Oscillator circuit design   265
Oscillators, sawtooth   60
O-pad   20, 23
Operating point   116, 140
Optics, fiber   90
Optics, phototransistor   81
Output admittance   36
Output power   151
Overall feedback   124
Overdriving   132

P

Pad   2, 20
Passive filter   1
Peaking filter   20
Phase inversion   133
Photometric system, phototransistors   85
Phototransistor   81
Phototransistor logic circuits   102
Pierce oscillator   269
Pi-filter   11
Pi-pad   23
Plastic transistors   156

Point light sources, phototransistors   87
Power-amplifier, RF   184
Power dissipation   168
Power output   150
Power supplies, amplifier   164
Power supply   168
Push-pull 119,   151

Q

Q-factor (RF)   28

R

Radiant energy   81
Radiation sensitivity, phototransistor   85
Radiation sources, phototransistor   85
Radiometric system, phototransistor   85
RC
  coupling   113
  filter design   4
  filters   2, 111
  oscillator   289
Refraction, fiber optics   91
Regenerative amplifier   121
Relaxation oscillator   291, 297
Resistive-coupled inverter   242
Resistor frequency limitations   109
Resonant circuit, oscillator   279
Resonant circuits (RF)   26
Resonant frequency   28
Reverse transadmittance   36
RF
  amplifier design   30, 170
  circuit design   26
  coils   30
  design   194
  mixer and converter   174
  multiplier   54, 184
  power amplifier   52, 184
  voltage amplifier   48, 170, 200
Rolloff, amplifier   115

S

Sawtooth oscillators   60
Schmitt trigger   72

Index

Secondary breakdown   146, 167
Series chopper   211
Series inverter   244
Shunt chopper   214
Single-ended output   166
Single transformer inverter   241
Solid-state light sources   96
Source bypass   128
Spectral response, phototransistor   84
Speed-up circuits, inverter   248
Stability factors (RF)   39
Stability RF   175
Stability versus gain   125
Stage feedback   123
Stage gain   172
Starting circuits, inverter   247
Stray impedances   112
Steady-state design, phototransistor   97
Stern k factor   39
Stern solution   40, 44
Sufficient feedback   127
Supply voltage   172
Switch, FET   215
Switches, transistor   207
Switching, phototransistor   100
Switching time, multivibrator   68
Symmetrical attenuator (pad)   2

T

T filter   8
Thermal runaway   159
Three-transistor amplifier   157
T-pad   22
T-pad, bridged   24
Transadmittance, forward   34
Transadmittance, reverse   36
Transducer gain   41
Transformer characteristics   151, 172
Transformer coupling   114
Transformer inverter   251

Transistor
　converter   235
　frequency limitations   108
　inverter   235
　oscillator   272
　switches   207
Trigger, Schmitt   72
Tuner, FM   203
Tungsten lamps, phototransistors   92
Tuning networks   48
Twin-T oscillator   287
Two-junction transistor amplifier   123
Two-junction transistor multivibrator   66
Two-port networks   30
Two-transformer inverter   239

U

UJT (unijunction transistor)
　multivibrator   70
　oscillator   297
　regenerative amplifier   121
　sawtooth oscillators   61

V

Variable-frequency oscillator   281
Vector algebra   32
Voltage amplifier, RF   200
Voltage variable capacitor (VVC)   54

W

Wave-generating circuit design   60
Worst case design   67

Y

y-parameter   32, 33
y-parameter measurements   36